Rudimentary Chemical Thermodynamics

RUDIMENTARY CHEMICAL THERMODYNAMICS

H. F. FRANZEN

ASSOCIATE PROFESSOR OF CHEMISTRY

B. C. GERSTEIN

ASSOCIATE PROFESSOR OF CHEMISTRY

BOTH OF IOWA STATE UNIVERSITY
AMES, IOWA

D. C. HEATH AND COMPANY

Lexington, Massachusetts

Published simultaneously in Canada.

Printed in the United States of America.

Library of Congress Catalog Card Number: 73-125128

Preface

There must necessarily be a strong justification for adding a thermodynamics text to a market already well supplied with material on this "closed" subject, and, indeed, with excellent texts at both advanced and elementary levels. The authors feel that such a justification exists for the present text in that it presents a difficult and abstract topic in simple and concrete terms through the use of examples and relatively simple mathematical formalism. While being no less rigorous than current "elementary" texts, the subject matter is nevertheless presented, we feel, in fewer pages at a more understandable level, i.e., at a "rudimentary" level.

The text was planned for use by students who after a year of high school chemistry are encountering their first and second years of college chemistry. We fully realized that an acquaintance with the elementary ideas involved in the application of differential and integral calculus is required for the understanding of the material as presented. We have, however, developed the treatment in such a way that complete familiarity with such applications is not required. Even at this level we have been surprised and pleased that our manuscript has become popular among graduate students studying for their preliminary examinations. Perhaps one reason for such a reception was the considerable influence of the styles of R. P. Feynman (in *The Feynman Lectures on Physics*, Addison-Wesley, 1963) and Henry A. Bent (both personally and in his delightful text *The Second Law*, Oxford, 1965) on our own method of presentation. As chemists actively pursuing research careers involving studies of species produced in the "high temperature" (ca. 2000°K) range correlated with thermodynamic information obtained in the 0.3–300°K range, we feel relaxed in our discussions of the second and third laws and have made a conscious effort to convey our own enjoyable experience with the material to the reader.

We would be less than truthful if we did not say that we are particularly enthusiastic about the treatment of the Second Law of Thermodynamics. It starts with facts that the student intuitively "knows." From there, it works, with particular attention to algebraic clarity, to a conclusion which is abstract but nevertheless given meaning by the connection between the student's intuition and the simplicity of concept. The development of the concept of temperature at this elementary level is also one that we have found to be practically and pedagogically useful.

The treatment, at the level of this text, of the conditions for thermodynamic equilibrium in heterogeneous systems has the advantage of being correct, and of utilizing a simple example to illustrate the principle at the "rudimentary" level. This is not to say, of course, that advanced treatments such as that found in Guggenheim's *Thermodynamics* are incorrect; they are simply inappropriate at our level of presentation.

The chapter on the Third Law of Thermodynamics and applications to high temperature phenomena was written by the one of us (H. F. F.) whose research program deals exactly with such calculations and the preparation of new compounds in the 1000–2000°K range. This chapter, we feel, is a prime example of our intent to make thermodynamics a subject which is applicable to urgent problems of modern chemistry.

There are two further aspects of this "rudimentary" text which make it, to our minds, a valuable contribution. The first is the emphasis of the use of the literature; we feel that this emphasis cannot be made too early. The second is the discussion of "standard states," a subject which, in our collective teaching experience, has been a particularly difficult one for students to absorb at both the elementary and the graduate levels. It is our feeling that our exposition of this much-misunderstood subject matter, so important to the intelligent use of thermodynamic data tabulations, is one which will be of value to students of thermodynamics as they work to apply its principles.

H. F. Franzen
B. C. Gerstein

Table of Contents

Chapter I *The First Law and Conservation of Energy*

Chapter II *The Second Law, A Criterion for Reversibility*

Chapter III *Chemical Equilibrium and the Second Law*

Chapter IV *Colligative Properties*

Chapter V *The Third Law and High Temperature Equilibrium Calculations*

Chapter VI *Electrochemistry and Chemical Equilibrium*

Chapter VII *Phase Diagrams*

Rudimentary Chemical Thermodynamics

CHAPTER I

The First Law and Conservation of Energy

THE ROLE OF THERMODYNAMICS

Recently there appeared a government bulletin warning chemists who work with finely divided zirconium of a safety hazard. The event responsible for this bulletin was an explosion resulting from rapid combination of finely divided Zr with O_2. There have been no recent safety bulletins warning chemists not to mix large quantities of H_2 and O_2, but then, *everyone* knows that these two elements form a mixture which happily explodes when "tickled" properly. How do we know? Some poor fellow many years ago probably lost his eyebrows and most of his hair finding out . . . through an explosion! It is one thing to recognize that, under given conditions, certain chemical reactions occur more or less readily and quite another to predict, on the basis of past experience, that such reactions are likely to occur again. As scientists and as human beings we clearly *do* make such predictions on the basis of past experience. It sometimes even turns out that scientists can more reliably make such predictions than the nonscientist. One of the disciplines which gives the chemist the edge over the nonscientist is known as chemical thermodynamics. This discipline enables the chemist to make the most reliable use of past experience in predicting the tendency for chemical reactions to occur.

The authors of this text believe that the most useful predictions regarding "chemical systems" are those to which numbers can be assigned. What do we mean by "chemical systems"? A **system** is generally defined as a region of matter which is under observation. A system has one or more characteristics called **properties** to which numbers can be assigned. For example, consider this text to be a "system." Some properties are its mass, volume, temperature, and the amount of energy which might be obtained from it if it were burned to generate steam which in turn pushed a piston. The properties mentioned are only a few possessed by the system. In fact it has an *infinite* number of properties! There is the volume V, and V^2, and V^3, ad infinitum. Of what use is it to talk about properties? The answer is that although we recognize that V^2, V^3, etc. are certainly properties, it is of little value to talk about them. In practice we limit ourselves to those properties which (experimentally, to be sure) are found to be most useful and then to the minimum number which will make the description of our physical surroundings easiest *and* most interesting. It will develop, of course, that some of the properties which have been found most useful are not at all those with which you are familiar; their *names* are not even familiar. In this context, it is important to realize that with this new subject, as with any new field of learning, a fair portion of the learning process is devoted to "learning the language." You have already started since you now know what *system* and *property* mean in our context, and you also know an example of a property, namely *volume*.

You undoubtedly will encounter other unfamiliar properties; for example, there are *internal energy*, *enthalpy*, *entropy*, *Gibbs free energy*, *free energy function*, and *heat capacity* as well as *temperature* for which you may have some physical feeling. We will later discuss types of "processes," such as *adiabatic* and *isothermal*, and the word *state* will frequently occur. Have no fear! You probably already know what is meant by these terms in a kind of vague, descriptive way. One of the purposes of this text is to make more precise the meanings of these concepts in a way that you will realize how past experience can be used to give you vast power in "chemical prophesy."

BASIC DEFINITIONS

Before we are able to talk meaningfully about the use of any of the above "properties," or "processes," we must clearly define *state* and *state of rest*. An example will help illustrate what we mean by these terms. Consider an unexploded mixture of hydrogen and oxygen.

There are many ways of describing this system. The late Welsh poet Dylan Thomas would no doubt have done it differently than we, and who is to say which is the better? We would point out that the *thermodynamic* method of describing a system is to associate numbers with properties, e.g., to specify numerically the volume, temperature, concentration, pressure, viscosity, refractive index, density, and magnetic susceptibility, information that is perhaps nonpoetic, but nevertheless useful. Such a description is said to specify the **state** of the system, for the *state* is *the collection of all properties.* If accurate measurement of the above properties yields the same result for an isolated system over long periods of time, the system is said to be in a particular kind of state, namely a "state of rest." **Thermodynamics deals only with states of rest, and changes in properties between states of rest.**

In our discussion of *property* thus far, we have discussed only the type of property about which we wish to talk, namely **macroscopic properties.** By macroscopic property is meant a property which is defined completely independently of the atomic nature of matter, and which, indeed, is undefinable in systems consisting of small collections of atoms. For example temperature, pressure, concentration (this one is less obvious), and magnetization are macroscopic properties. On the other hand, the average velocity of molecules in a large system is a property of the macroscopic system, but not a macroscopic property. Classical thermodynamics deals only with macroscopic properties.

NOW we can say what we mean by **thermodynamics.** It is the discipline which enables us to apply our relevant experience with regard to energy exchange and spontaneity to any process which involves a change in a macroscopic property of a system between two states of rest. You are still being cheated a little, however, because as yet we have not discussed **energy** in the way that we would like to have you understand the term. This will be taken care of, however, in the following discussion.

WORK

The thermodynamic concept of work is the same as that with which you first came in contact in your high school physics course, namely **work** is a force operating along a distance. (It should be noted here that the meaning of *force* and *distance* in electrical and magnetic work are subject to some degree of ambiguity. We shall for the present ignore these types of work.)

The standard, albeit dry, example of thermodynamic work is that done on a piston. Consider a piston weighing, say, 200 kg in a cylinder (see Fig. 1.1), containing a gas at some pressure P. Such a

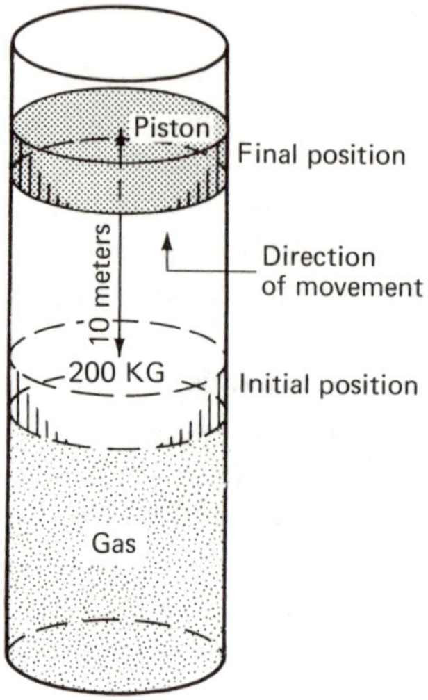

Figure 1.1 Thermodynamic work illustrated as the movement of a force through a distance.

system might be a component of a rather large engine. Suppose that the pressure of the gas (not necessarily uniform throughout the cylinder during a change in the position of the piston) is sufficiently large to raise the piston. Let the piston move ten meters upward.

The downward force exerted by the piston is Mg, where M is the mass of the piston and g is the acceleration due to gravity, 980 cm/sec^2 at sea level. The work done by the gas on the piston in raising the piston a distance 1,000 cm is

$$\begin{aligned} w = Mgh &= 200{,}000 \text{ g} \times 980 \text{ cm/sec}^2 \times 1{,}000 \text{ cm} \\ &= 19.6 \times 10^{10} \text{ erg} \qquad \textbf{(I-1)} \\ &= 19.6 \times 10^{10} \text{ erg} \times 10^{-7} \text{ joules/erg} \\ &= 19.6 \times 10^{3} \text{ joules} \end{aligned}$$

If we now introduce the conversion factor 4.184 joules/calorie, a unit we shall define later in this chapter, we obtain for the work done on the piston 4.45×10^3 calories, or 4.45 kilocalories.

We can obtain a slightly different form for the work equation by recalling the definition of pressure, i.e.,

$$P = F/A \tag{I-2}$$

the force divided by the area. For the system of Fig. 1.1 the work done by the system is

$$w = F_e(h_2 - h_1) \tag{I-3}$$

where F_e is the external force on the system (Mg in this case) and h_2 and h_1 are the final and initial heights of the piston. Multiplying and dividing by A, the area of the piston,

$$w = \frac{F_e}{A}(Ah_2 - Ah_1)$$

$$= P_e(V_2 - V_1) = P_e \Delta V \tag{I-4}$$

i.e., the work done by the system is the external pressure times the volume change of the system. This relation applies to the portion of the work done by the system between any two heights during the overall process. We may compute the total work as the sum

$w = P\Delta V_1 + P\Delta V_2 + \cdots + P\Delta V_n$ (where n is the total number of steps).

We write this

$$w = \sum_{i=1}^{n} P\Delta V_i \tag{I-5}$$

This may graphically be represented as in Fig. 1.2, where it is clear

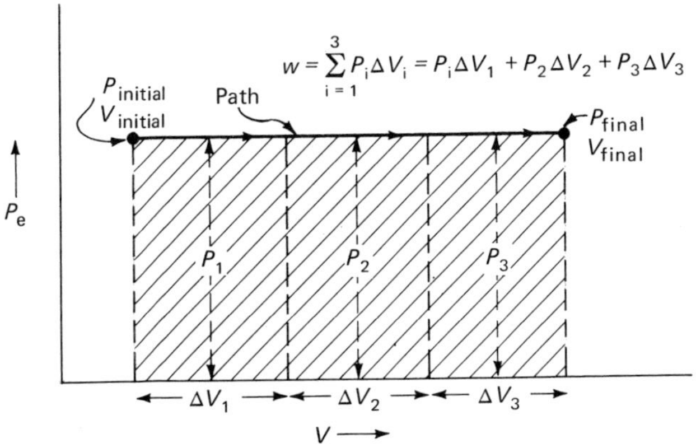

Figure 1.2 Graphical representation of work as the area under the P_e–V curve for a constant pressure path.

that w is the total area under the P_eV curve, i.e., $P\,\Delta V_{\text{total}} = P(V_{\text{final}} - V_{\text{initial}})$. All of the above discussion has applied to a constant external pressure. If the external pressure had varied with each step, the appropriate sum would be

$$w = \sum_i P_i\,\Delta V_i \tag{1-6}$$

where P_i is the external pressure during the volume change ΔV_i. The work may now be represented graphically as shown in Fig. 1.3, where

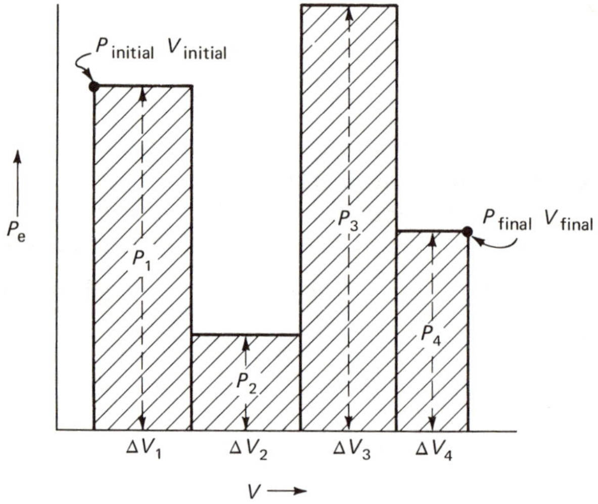

Figure 1.3 Work as the area under the P_e–V curve for the case of a variable pressure process.

again the work is the area under the $P_e - V$ curve. Now if the truth be known, the variable pressure shown in Fig. 1.3 is pretty silly, and for the most part, one does not find such behavior in physical systems.

In general, one might expect a rather smooth pressure-volume relationship such as shown in Fig. 1.4. Again, the work done by the system is given by the area under the $P_e - V$ curve. This may be seen by imagining the process to occur in a series of constant-pressure steps, as in Fig. 1.5, and illustrated by the solid lines in the figure. Of course, as Fig. 1.5 is constructed, the stepwise path giving us an approximation to the true path is clearly a very rough approximation to the true path, and the area of the rectangles is clearly a very rough approximation to the area under the true curve. We can make a

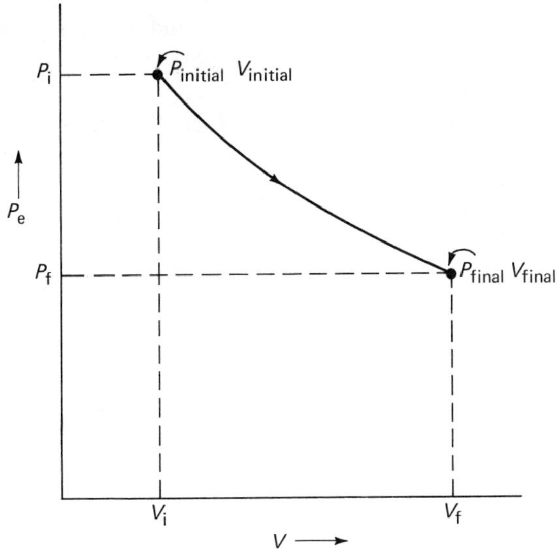

Figure 1.4 Typical smooth P_e–V curve for a system doing pressure-volume work.

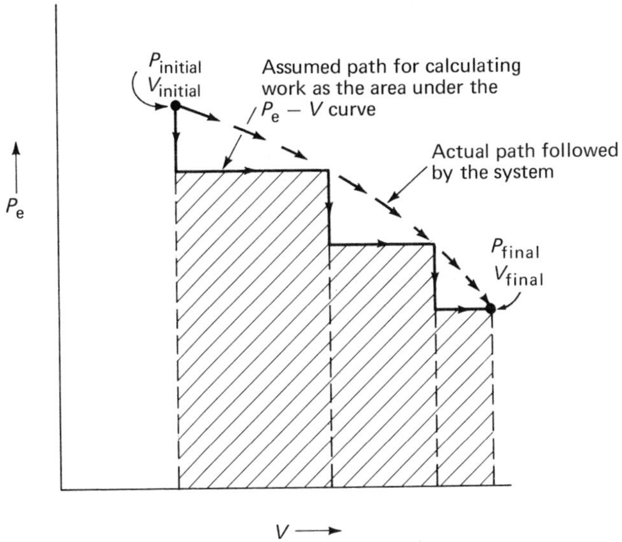

Figure 1.5 Approximation of smooth P_e–V curve by constant-pressure, constant-volume steps.

better approximation by taking smaller and smaller pressure steps, thus increasing the number of rectangles. Clearly, the greater the number of steps we take, the closer we approach the true curve. By taking an increasingly large number of smaller and smaller steps, we can approximate the smooth curve to any degree of accuracy desired. Using the previously developed symbolism, we can write the first "four step" approximation for the work as

$$w \cong \sum_{\mathrm{i}=1}^{4} P_{\mathrm{i}}\, \Delta V_{\mathrm{i}} \tag{I-7}$$

where $\cong$ indicates an approximate equality. For the limit of the process of taking an increasing number of smaller and smaller steps we can write the equation

$$w = \lim_{n \to \infty} \sum_{\mathrm{i}=1}^{n} P_{\mathrm{i}}\, \Delta V_{\mathrm{i}} \tag{I-8}$$

which means "the summation of $P_{\mathrm{i}}\, \Delta V_{\mathrm{i}}$ in the limit as the number of steps becomes very large (i.e., infinite)." Since we have increased the number of steps sufficiently to give the "true curve" to any desired degree of accuracy, we write the equation as an equality rather than an approximate equality. A simpler way of writing this equation is to use a long, thin S to represent the infinite summation, and dV to represent all the infinitesimal ΔV's.

$$w = \int_{V_{\mathrm{initial}}}^{V_{\mathrm{final}}} P\, dV \tag{I-9}$$

You will recall from the integral calculus that this symbolism means "the integral of $P\, dV$," and if P is expressed as a known function of V you will be able to perform the integration either by remembering the integral in question or by consulting tables. It is important to note that in our discussion of $P - V$ work, we have talked about work done *by* the system which is $\int P_{\mathrm{e}}\, dV$. Thus if the system expands at constant pressure, as in our above example, V_{final} is greater than V_{initial}, and $w_{\text{by the system}} = P_{\mathrm{e}}(V_{\mathrm{final}} - V_{\mathrm{initial}})$, a positive quantity. The work *on* the system, of course, is just the negative of the work *by* the system; i.e.,

$$w_{\text{on system}} = -\int P_{\mathrm{e}}\, dV \tag{I-10}$$

By convention, we will always speak of work as being work done *on* the system. This is to say, for $P - V$ work in particular, we write

$$w = -\int P_e \, dV \tag{I-11}$$

Before continuing our discussion of work it is necessary to make a few remarks about temperature, a property about which thermodynamics has a great deal to say.

In order to proceed we must recognize that a property, i.e., a number, can be associated with the tendency for heat to flow from one body to another. (We assume at this point that you know qualitatively what heat is, although we define heat at a later point in our discussion). If we place two chunks of metal in contact, and if one "heats" and the other "cools" (as determined by, say, feeling the chunks), then we say that the one that "cooled" had a higher temperature than the other. If, in fact, no change is detected, we say that the two are in thermal equilibrium, or are at the same "temperature."

If we can find a system with a property such that its numerical value allows us to determine whether or not the system will be in thermal equilibrium with another system, we will have found an *empirical thermometer*. There are, in fact, many such empirical thermometers in use: the length of a mercury column, the resistance of a platinum wire, the emf of a thermocouple, the brightness of a tungsten filament, and so on. One very useful thermometer is the gas thermometer which yields, for the limiting case of infinitely dilute gases, a temperature proportional to the pressure-volume product, a property of the gases.

For the remainder of this chapter and the early part of the next we shall use the word "*temperature*" to mean this "*gas thermometer*" *temperature*. We shall, in a later chapter, complete our discussion of the Second Law of Thermodynamics by defining the thermodynamic temperature. We shall give that temperature the same symbol as the one used with the gas thermometer, for the two temperatures have the same numerical values although they are defined in different ways. On the thermodynamic scale, with the Celsius degree chosen as the size of the degree, the temperatures of the melting and boiling points of water are roughly 273°K (Kelvin) and 373°K, respectively. Room temperature is generally taken to be 298°K, and thus most thermodynamic data are tabulated with this as a reference temperature.

For calculations which involve differences in temperatures (i.e., first law calculations) only the size of the degree is important, not the zero. Thus for many purposes it is common practice to use

Celsius temperatures. We have followed this practice in the example at the end of this chapter.

An interesting fact which has been observed to be true for pure substances in states of rest and in the absence of external fields such as electrical, gravitational, and magnetic fields is that there is a functional relationship between T, the empirical temperature of the substance; P, the pressure exerted by the substance; n, the number of moles of substance; and V, the volume occupied by the substance. We can write

$$P = f(\mathrm{T}, V, n) \tag{I-12}$$

i.e., the pressure exerted by the substance in a state of rest is a function of the temperature, volume and number of moles of the substance. One of the simpler examples of such an *equation of state* is that of an "ideal gas," for which

$$P = n\mathrm{RT}/V \tag{I-13}$$

where R is a constant. This can be shown *via* kinetic theory to be the equation of state of a collection of noninteracting particles. We shall have more to say about the relationship of this equation of state to the equations of state of real gases in Chapter II.

At this point we wish to illustrate the utility of equations of state with regard to the calculation of reversible work. We shall use the ideal gas equation of state as an example. Let the external pressure on a system be denoted by P_{ex} and the internal pressure exerted by the system be denoted by P_{int}. During a *reversible* process $P_{\mathrm{ex}} = P_{\mathrm{int}}$, and P_{int} is given by the equation of state. Thus

$$\begin{aligned} w^{\mathrm{rev.}} &= -\int P_{\mathrm{ex}}\, dV \\ &= -\int P_{\mathrm{int}}\, dV \\ &= -\int \frac{n\mathrm{RT}}{V}\, dV \end{aligned} \tag{I-14}$$

for an ideal gas. For a *general* reversible process the work is the integral of the pressure given by the equation of state times the differential of the volume. We note that in the case of a volume change of an ideal gas at constant temperature this becomes

$$w^{\mathrm{rev.}} = -\int_{V_1}^{V_2} \frac{n\mathrm{RT}}{V}\, dV$$

$$= -n\text{RT} \int_{V_1}^{V_2} \frac{dV}{V} \qquad \text{(I-15)}$$
$$= -n\text{RT} \ln V_2/V_1$$

The relationship of work to the properties of thermodynamic systems is: a change in the properties of a system can be effected by the influence of work. Years of observation have shown that, *in general*, the amount of work necessary to change a system from one given state to another depends upon HOW the work is done. It has also been observed that the amount of work necessary to bring about a specified change of state in a system *isolated* from all external influences such as light radiation and vibration (*except* the influence effecting work on the system) is always the same, regardless of how the work is done on the isolated system. For example, a weight falling through a distance and doing work by turning a paddle wheel in an isolated water bath will heat the water just as much as an equivalent amount of electrical work being introduced into the bath by a current passing through a resistance element in the bath. A system so isolated is called an **adiabatically isolated system.**

THE FIRST LAW OF THERMODYNAMICS

The fact that adiabatic work is independent of path is a statement of the First Law of Thermodynamics. This statement, in terms used earlier, may be rephrased as follows: "Adiabatic work is a measure of change in a *property* of a system." We call this property **Internal Energy,** and give it the symbol U. Thus we write

$$U_2 - U_1 = \Delta U = w_{\text{adiabatic}} \qquad \text{(I-16)}$$

where U_2 is the internal energy in the final state, and U_1 is the internal energy in the initial state. Note carefully that we have defined only *changes* in U, and that the zero of U is in general undefined. We can, nonetheless, meaningfully discuss the property U leaving the zero unspecified, and in fact experience has shown that U, even with its zero undefined, is a particularly valuable property to investigate, in the sense that we mentioned in discussing *properties* on page 3.

HEAT

One of the first uses to which the internal energy is logically put is in the formulation of the definition of heat. We can investigate the change in internal energy in a system which is not adiabatically isolated. We can also measure the work as previously described. In general, the two are not the same, and we define the heat, q, as the difference between ΔU and w, i.e.,

$$\Delta U = q + w \tag{I-17}$$

This is the general statement of the First Law of Thermodynamics, and is simply a formal accounting of energy changes in terms of heat and work. For the general change in a system, $q + w$ is therefore a property change. It should be very clear that while $q + w$ is a property change, neither q nor w are in general property changes, except under conditions where one or the other is zero. This is to say that for a given change in state of a system, neither the heat added to the system nor the work done on the system is uniquely determined. There are in general an infinite number of ways of adding heat and doing work between the same initial and final states. It is only the *sum* of the heat and the work that is the same between given initial and final states! For example, consider 0.1 mole of an ideal gas in a state described by specifying the volume to be 1 liter and the pressure to be one atmosphere. Let this system be adiabatically and reversibly compressed to a state having a volume of 0.5 liter and a pressure of 3.17 atm. The path taken by the system in the above described process is shown in Fig. 1.6 as the solid line labeled "Path 1," and the work, of course, is the area under the $P - V$ curve, shown as the darkened area. For an adiabatic process, $\Delta U = w =$ the area under the curve and $q = 0$. The area under the curve corresponds to 21.3 cal. If we now took the same system from the same initial state to the same final state, but along the path indicated by the dashed lines in Fig. 1.6, ΔU, being a property change, would be the same as before. The work, however, is again given by the area under the $P - V$ curve, and for the path shown is $P_{\text{initial}}\,\Delta V$ (since the constant volume part of the path does not contribute to the area). This work can be shown to be 12.2 calories (the student may evaluate this for himself by calculating the area under the $P - V$ rectangle for Path 2, and converting from liter-atmospheres to calories by using the conversion formula (liter $-$ atm) $\times \dfrac{1.986}{0.0820} =$ cal).

Since ΔU is the same for both paths, q for the second path is 9.1 cal, so that

$$\begin{aligned} \Delta U &= q + w \\ \text{Path 1: } 21.3 &= 0 + 21.3 \text{ cal} \\ \text{Path 2: } 21.3 &= 9.1 + 12.2 \text{ cal} \end{aligned} \tag{I-18}$$

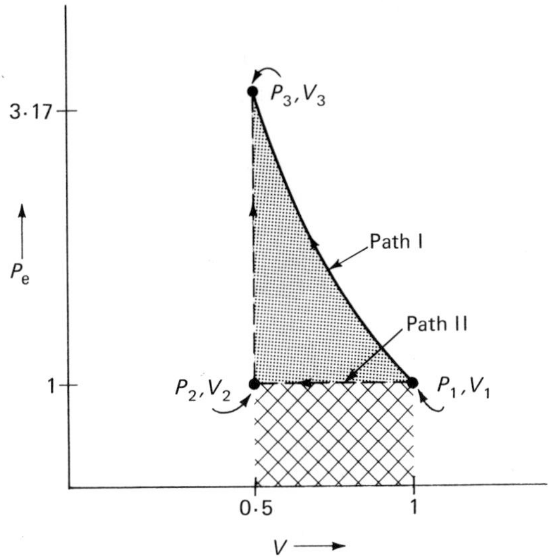

Figure 1.6 Illustration of work as being path dependent. For Path I, from P_1, V_1 to P_3, V_3, work (w) is the shaded plus crosshatched areas. For Path II, from P_1, V_1 to P_2, V_2 and then to P_3, V_3, w = crosshatched area.

It is clear that any intermediate path between Paths 1 and 2 would be a possible path, and would involve different values of both q and w.

HEAT CAPACITY

It is a matter of common experience that the difference between hot and cold coffee is generally related to a source of heat, i.e., we increase the temperature of a cup of cold coffee by heating it! This is so mundane as to be almost trivial, but the process of increasing temperature by heating is one which nevertheless is related to an important property, namely heat capacity at constant volume or at constant pressure. This is just the ratio of added heat to temperature rise, the heat being added along either a constant volume or along a constant pressure path, in the limit as the temperature interval becomes vanishingly small. This formidable statement may be formalized as follows:

$$C_p = \lim_{\Delta T \to 0} \frac{q_p}{\Delta T} \tag{I-19}$$

$$C_v = \lim_{\Delta T \to 0} \frac{q_v}{\Delta T} \tag{I-20}$$

The subscript p in equation (I-19) and the subscript v in equation (I-20) indicate that the heat is introduced along paths of constant P and constant V, respectively. One of the principal uses of the heat capacity of a system is the determination of the amount of heat required to raise the temperature of the system by a prescribed amount. The amount of heat introduced into a system is, in general, dependent upon the path through which the system passes as the heat is introduced. We write

$$q_{\text{path}} = \int_{T_{\text{initial}}}^{T_{\text{final}}} C_{\text{path}}\, dT \tag{I-21}$$

where the integral sign, as before, is taken to mean the appropriate sum over a large number of terms. Two easily obtainable experimental paths are constant volume and constant pressure paths. For the constant volume case, we write, for example,

$$q_v = \int_{T_i}^{T_f} C_v\, dT \tag{I-22}$$

As an example of the use of heat capacities, consider the following. It is well known that the reaction of N_2(gas) with H_2(gas) to form NH_3(gas) proceeds sufficiently rapid to be industrially feasible only at temperatures exceeding 500°C. Let us calculate the amount of heat required to raise the stoichiometric amounts of N_2(g) and H_2(g) required for the formation of one mole of NH_3(g) from 25°C to 500°C. For the purposes of this example, we shall take the molar heat capacity of each reactant to be constant over the temperature range in question, and equal to 7 cal/(mole deg). This value is an approximation which will give an answer accurate to $\frac{1}{2}\%$. Since $\frac{1}{2}$ mole of N_2(g) and $\frac{3}{2}$ moles of H_2(g) are needed to form one mole of NH_3(g), the heat capacity of the mixture is the sum of $\frac{1}{2}$ the molar heat capacity of N_2(g) and $\frac{3}{2}$ the molar heat capacity of H_2(g); i.e., $C_p(\text{mixture}) = \frac{1}{2}\overline{C}_p(N_2) + \frac{3}{2}\overline{C}_p(H_2) = 14$ cal/deg, where the symbols $\overline{C}_p$ denote molar heat capacities at constant pressure, and we have assumed that the mixture is heated at constant pressure (i.e., with increasing volume) in this problem. Since the heat capacity is, by assumption, constant over the temperature range, the total heat added to raise the temperature by 475 Celsius degrees is 14 cal/deg $\times$ 475 deg $=$ 6,650 calories.

UNITS OF ENERGY

A number of different units for expressing amounts of energy have been mentioned thus far, namely ergs, liter-atmospheres, joules and calories, and the conversion from one unit to another has been treated (for example in equation I-1). The definition of the erg (dyne-cm) and the liter-atm are mechanical in nature and follow naturally from the definition of work as force times distance. The definition of the joule is electrical in nature. The joule is the energy associated with the drop of one coulomb of charge through one volt. Joules are measured by measurements of currents, volts and time. The unit of energy commonly used in discussions of chemical thermodynamics is the calorie. The calorie was in the past defined as the amount of heat required to raise the temperature of one gram of water from 14.5° to 15.5°C. Currently, the joule is taken to be the standard unit and the calorie is defined to be exactly 4.1840 joules. A reason for this definition is that the device used to measure quantities of heat, the calorimeter, is usually calibrated electrically—a precisely measured current is passed through a precisely measured voltage drop for a precisely measured period of time such that the energy goes to heat the calorimeter, and the temperature rise of the calorimeter is measured. Thus the calorimeter is calibrated so that the scientist knows how much energy is required to increase the temperature of the calorimeter by a given amount.

Another unit of energy commonly used by chemists, and frequently used throughout this text, is the kilocalorie (abbreviated kcal), which is one thousand calories.

THE FIRST LAW AND CHEMISTRY

Having dealt at some length with the concepts of internal energy, heat and work, and heat capacity, all of which seem at first rather nonchemical, we are finally in a position to discuss the power of thermodynamics as applied to problems of chemical importance.

A particularly important concept in chemistry is that of the chemical bond. To the extent that individual bonds exist in molecules, we can assign certain characteristic "strengths" or, in common terminology, "bond energies" to bonds in molecules, and use these energies to test theories of the chemical bond. From this viewpoint we are able to generally understand a great deal of chemistry. A case in point is the carbon-hydrogen bond energy. We all have a vested interest in this bond inasmuch as compounds containing carbon and hydrogen form a large proportion of our physiological makeup. To determine the energy of this bond consider the reaction

$$C(s) + 2\,H_2(g) = CH_4(g) \tag{I-23}$$

The average C—H bond energy in methane can be calculated from the heat absorbed during this reaction, providing the energy of sublimation of carbon, i.e., of the reaction

$$C(s) = C(g) \tag{I-24}$$

and the energy of dissociation of hydrogen, i.e., of the reaction

$$H_2(g) = 2\,H(g) \tag{I-25}$$

are known. Even with the energy changes of reactions (I-24) and (I-25), which are known, we would be at a loss to obtain the energy of the C—H bond in methane if we had to directly determine the heat absorbed in reaction (I-23). We simply are not able to conduct an experiment in which one mole of carbon reacts with two moles of hydrogen to form one mole of methane and the heat absorbed during this reaction is measured. However, with relative ease, we can measure the heats developed when each of the reactants and products is burned in an oxygen atmosphere. The reactions which occur when the reactants and products of equation (I-23) are burned and the heats absorbed, q_v, at constant volume are

$$C(s) + O_2(g) = CO_2(g) - q_v \tag{I-26}$$

$$H_2(g) + \tfrac{1}{2}\,O_2(g) = H_2O(l) - q'_v \tag{I-27}$$

$$CH_4(g) + 2\,O_2(g) = CO_2(g) + 2\,H_2O(l) - q''_v \tag{I-28}$$

When such heats are measured, as they are by combustion calorimetry, the reactants and products are usually at 25°C. If we were so fortunate as to be allowed to add and subtract equations, such as equations (I-26), (I-27), and (I-28), with the corresponding heats being additive, we could calculate the energy change for the carbon-hydrogen reaction which produces methane as follows: [reaction (I-26)] + 2 × [reaction (I-27)] − [reaction (I-28)], which yields

$$C(s) + 2\,H_2(g) = CH_4(g) - q'''_v \tag{I-29}$$

The plain facts of the matter are that we *are* allowed to add reactions algebraically, and obtain the heat of the summed reactions by algebraically summing the heats of the individual reactions, if we do so as outlined below. The heat of reaction (I-29), q'''_v, is equal to q_v +

$2q'_v - q''_v$ and is the heat absorbed when one mole of methane is formed from the elements at constant volume and temperature (i.e., the initial and final temperatures are equal). This is true because the heat absorbed at constant volume is the change of a property of the system, and is therefore independent of path. We can accomplish the synthesis of methane from the elements, according to equation (I-23), in three steps. The heat absorbed in each of the steps can be experimentally determined *via* combustion calorimetry. The three steps are: (1) reaction (I-26), (2) reaction (I-27) twice, (3) reaction (I-28) in reverse.

The fact that the heat absorbed at constant volume is a property change follows from the First Law of Thermodynamics. The change in internal energy is equal to the sum of the heat absorbed by the system and the work done on the system. But at constant volume no work is done on the system (dV is zero), and thus q_v is equal to the change in internal energy, which is certainly a change in a property, i.e., is independent of path.

We now begin to reap the benefits of the above formalism regarding such relatively nonchemical concepts as heat and work. By simply realizing that q_v is a property change we are in a position to determine heats evolved or absorbed in reactions we have no hope of carrying out quantitatively. We obtain these heats simply by summing heats for appropriate reactions which can be easily caused to go. In this way we are able to obtain a vast amount of information regarding strengths of chemical bonds and energy relationships for processes from a relatively small number of experiments.

In Table I.1, for example, the energies required to dissociate methylidyne, CH, methylene, CH_2, methyl, CH_3, and methane, CH_4, to the

Table I.1

Changes in Internal Energy for Various C—H Dissociation Reactions

Reaction	$\Delta U_0^\circ = q_v$ at 0°K kcal
$CH(g) = C(g) + H(g)$	80
$CH_2(g) = C(g) + 2\,H(g)$	178
$CH_3(g) = C(g) + 3\,H(g)$	292
$CH_4(g) = C(g) + 4\,H(g)$	392

gaseous atoms are tabulated. The temperature to which these values correspond is 0°K (−273°C); the reasons for choosing this temperature will be developed in later chapters. The tabulated energy changes indicate the strengths of the bonds in the compounds. The average C—H bond energy increases from 80 kcal in CH to 392/4 = 98 kcal in CH_4. It should be noticed that while the average bond energy in CH_4 is 98 kcal, the energy required to strip one hydrogen atom from CH_4 is 392 − 292 = 100 kcal, which is a slightly higher value.

ENTHALPY

Unfortunately most chemists do not work with systems at constant volume. For practical purposes we work in a constant pressure environment, however. In line with our initial attack on useful and interesting properties, we now look for a measurable property suitable to our laboratory conditions; namely, we ask whether the heat evolved or absorbed during a process which occurs at constant pressure is a change in some property. The answer is "yes," and this can be seen as follows. We know that U, P and V are properties. Therefore the quantity

$$H = U + PV \tag{I-30}$$

is also a property. At constant pressure, then

$$\Delta H = H_2 - H_1 = (U_2 - U_1) + P(V_2 - V_1) = \Delta U + P\Delta V \tag{I-31}$$

where P is the constant pressure at which the process occurs. We also know that at constant pressure,

$$\Delta U = q_p - P\Delta V \tag{I-32}$$

Therefore, at constant pressure,

$$\Delta H = q_p - P\Delta V + P\Delta V = q_p \tag{I-33}$$

and q_p (which is equal to ΔH) is a property change. The property, H, is called the **enthalpy,** which is a word related to the Greek verb meaning "to heat." It should always be kept in mind that the utility of H results from the fact that at constant pressure, ΔH = the heat added to a system during a process.

There exist tables of enthalpy changes for chemical reactions, such as Table I.2, which allow us to calculate the enthalpy change for a wide variety of chemical reactions. The tables are such that we are able to add reactions and enthalpy changes in order to obtain the

Table I.2

Enthalpy Changes for Chemical Reactions at 25°C, Reactants and Products in their Standard States

Reaction	ΔH° (kcal)
$H_2(g) + \frac{1}{2} O_2(g) = H_2O(l)$	−68.3
$C(s) + 2\,H_2(g) = CH_4(g)$	−17.9
$C(s) = C(g)$	170.9
$H_2(g) = 2\,H(g)$	52.1
$Na(s) + \frac{1}{2} Cl_2(g) = NaCl(s)$	−43.4
$\frac{1}{2} N_2(g) + \frac{3}{2} H_2(g) = NH_3(g)$	−11.0
$2\,Al(s) + \frac{3}{2} O_2(g) = Al_2O_3(s)$	−400.4
$Zr(s) + O_2(g) = ZrO_2(s)$	−245.5
$W(s) + 3\,F_2(g) = WF_6(g)$	−399.3

enthalpy change for a reaction of interest. The principle behind the use of the tables is exactly the same as the principle that was used to obtain the heat of reaction at constant volume for the formation of methane from the elements.

We shall illustrate the use of such tables with an example. Suppose that we wish to calculate the heat of reaction (enthalpy changes accompanying reactions are commonly referred to as **heats of reaction**) for the process

$$6\,CH_4(g) = C_6H_6(l) + 9\,H_2(g) \qquad \text{(I-34)}$$

the products and reactants to be taken at 25°C and one atmosphere applied pressure.* This reaction, the production of benzene from methane, is one which, as far as we are aware, has not yet been carried out. One important factor in determining whether or not the reaction can be made to occur, however, is the enthalpy change for the reaction. We have available the enthalpy changes for the reactions whereby all of the species in the above reaction are formed from their elements (these enthalpy changes have been obtained by burning substances and obtaining the heats absorbed, and then adding reac-

* We are making an approximation in taking the state of the gas corresponding to the tabulated values to be one atmosphere pressure. The nature of this approximation will become apparent to the reader in the next section, and it will also become apparent that the error introduced by making this approximation is negligible.

tions and heats exactly as in the methane illustration):

$$C(s) + 2\,H_2(g) = CH_4(g) \qquad \Delta H^\circ = -17.89 \text{ kcal} \tag{I-35}$$

$$6\,C(s) + 3\,H_2(g) = C_6H_6(l) \qquad \Delta H^\circ = 12.48 \text{ kcal} \tag{I-36}$$

With these values we determine the enthalpy change for reaction (I-34): $\Delta H^\circ = 12.48 + 6 \times 17.89 = 119.82$ kcal. This is to say that for each mole of benzene produced according to reaction (I-34), with the reactants and products at 25°C and one atmosphere pressure, the system *absorbs* 119.82 kcal.

STANDARD STATES

There is one small detail with regard to notation that must yet be clarified. Note that in equations (I-35) and (I-36) we have placed a small circle as a superscript on the symbols for the enthalpies of formation, and also on the symbol for the resulting enthalpy change for the formation of benzene from methane. This superscript indicates that the enthalpy changes are "standard" enthalpy changes, i.e., the reactants and products are taken to be in their **standard states. The standard state of a pure element or compound in a condensed state of aggregation is defined as the pure element or compound at the temperature in question and one atmosphere applied pressure. The standard state of a gas is the hypothetical ideal gas at one atmosphere pressure.** The hypothetical ideal gas standard state is the state in which the gas would be if it were compressed from very low pressure to one atmosphere and if it obeyed the ideal gas equation of state during the compression. It is hypothetical for two reasons. First, the gas in general will not behave ideally; second, in many cases (e.g., H_2O at 25°C) the real gas would condense before compression to one atmosphere. As we shall soon show, it is a simple matter to calculate the enthalpy change for the process of compressing the gas along the hypothetical path. The small approximation mentioned in the previous footnote on p. 19 is just due to the differences between the enthalpies of real gases at one atmosphere pressure and those of real gases in their standard states.

For example, the standard state of liquid water is liquid water at one atmosphere applied pressure (**not one atmosphere vapor pressure**), and the temperature of interest. The standard state of gaseous water, on the other hand, is hypothetically ideal gaseous water at one atmosphere pressure and the temperature of interest. If the temperature is sufficiently high the hypothetical ideal gas standard state and the state of the real gas at one atmosphere are **essentially** the same.

The *standard heat of formation*, given the symbol ΔH°_f, of a compound or element is the enthalpy change for the process: elements (or element) in reference states → compound (or element) in standard state. The reference states for the elements are the standard states of the elements in an arbitrarily chosen form, e.g., H_2(g) for hydrogen, solid graphite for carbon, liquid bromine, all at room temperature. The heats of reaction given in Table I.2 are all standard heats of formation of the compounds appearing in the given reactions.

It might be mentioned that if there is one thing that is standard about standard heats of formation and standard states, it is the confusion that they cause in the minds of students. They are important bookkeeping devices, however, and for that reason the preceding paragraphs have been, although dry, pertinent and ultimately useful.

TEMPERATURE DEPENDENCE OF ENTHALPY

The above calculations relating enthalpies of formation to heats of general reactions have one obvious limitation, namely, the standard heats of formation are generally tabulated at a reference temperature of 25°C. This means that unless we have a convenient method of determining how enthalpies vary with temperature we can only make heat of reaction calculations in which the products and reactants are at 25°C (the system may get hotter or colder during the reaction, but the initial and final temperatures must be 25°C). Surely very many important processes occur in which the reactants and products are at temperatures vastly different from 25°C. A case in point is the production of pig iron from iron ore, in which the following net reactions are important:

$$Fe_2O_3(s) + 3\,C(s) = 2\,Fe(l) + 3\,CO(g) \qquad \textbf{(I-37)}$$

and

$$CaCO_3(s) + SiO_2(s) + C(s) = CaSiO_3 + CO(g) \qquad \textbf{(I-38)}$$

These reactions occur at various temperatures in the blast furnace, the highest of which must be sufficient to melt both the iron and the calcium silicate. This is necessary in order that these products will flow to the bottom of the furnace, where the iron is drained off to form "pigs." The temperature required is about 1500°C. Thus, if we wish to make a calculation of the heats of reactions pertinent to the formation of pig iron at the temperatures at which the reactants are introduced and the products are formed in liquid form, it is necessary to understand how the enthalpy changes of the reactions vary as functions of temperature.

It is clear that at any temperature, the heat of reaction at constant pressure is just the total enthalpy of the products minus the total enthalpy of the reactants, both at that temperature. If we know how the enthalpies of the individual products and reactants behave as the temperature is varied, we can utilize the following relationship:

$$\Delta H_{T_2} = \Delta H_{T_1} + \sum_{\text{products}} (H_{T_2} - H_{T_1}) - \sum_{\text{reactants}} (H_{T_2} - H_{T_1}) \qquad \textbf{(I-39)}$$

This relationship is an immediate consequence of the fact that enthalpy is a property, and hence the enthalpy change around the closed path: (1) reaction at T_1, products heated to T_2; (2) reverse reaction at T_2, reactants cooled to T_1, is zero.

Consider the first net reaction given above, equation (I-37). To facilitate our discussion of the calculation of the heat of reaction at, say 1550°C, we have collected some enthalpy data pertinent to this reaction in Table I.3.

Table I.3

Enthalpy Data Pertinent to the Reaction of $C(s)$ *with* $Fe_2O_3(s)$

	Fe_2O_3	C	Fe	CO
$\Delta H_f°$, 25°C kcal/mole	−197.3	0	0	−26.4
$(\overline{H}^\circ_{1550°C} - \overline{H}^\circ_{25°C})$ kcal/mole	52.31	7.4	17.8	12.0

Note that the heats of formation of solid C and Fe at 25°C are zero, as required by the definition of heat of formation. Briefly postponing inquiry into the method used to obtain the numbers in the second row of Table I.3, let us find the enthalpy change for reaction (I-37) at 1550°C.

$$\begin{aligned}\Delta H^\circ_{1550°C} &= 3 \times (-26.41 \text{ kcal}) + 197.3 \text{ kcal} + (3 \times 12 \text{ kcal} \\ &\quad + 2 \times 17.8 \text{ kcal}) - (52.3 \text{ kcal} + 3 \times 7.4 \text{ kcal}) \qquad \textbf{(I-40)} \\ &= 115.2 \text{ kcal}\end{aligned}$$

This equation can be symbolically represented by:

$$\begin{aligned}\Delta H^\circ_{1550^\circ\mathrm{C}} = {} & 3\,\Delta H^\circ_{f,25\mathrm{C}^\circ}(\mathrm{CO}) - \Delta H^\circ_{f,25^\circ\mathrm{C}}(\mathrm{Fe_2O_3}) \\ & + [3 \times \{\overline{H}^\circ_{1550^\circ\mathrm{C}}(\mathrm{CO}) - \overline{H}^\circ_{25^\circ\mathrm{C}}(\mathrm{CO})\} \\ & \qquad + 2 \times \{\overline{H}^\circ_{1550^\circ\mathrm{C}}(\mathrm{Fe}) - \overline{H}^\circ_{25^\circ\mathrm{C}}(\mathrm{Fe})\}] \\ & - [\{\overline{H}^\circ_{1550^\circ\mathrm{C}}(\mathrm{Fe_2O_3}) - \overline{H}^\circ_{25^\circ\mathrm{C}}(\mathrm{Fe_2O_3})\} \\ & \qquad + 3 \times \{\overline{H}^\circ_{1550^\circ\mathrm{C}}(\mathrm{C}) - \overline{H}^\circ_{25^\circ\mathrm{C}}(\mathrm{C})\}]\end{aligned} \tag{I-41}$$

There is one point which is immediately obvious about the above symbolism, namely the complexity of its appearance. The meaning underlying the complex symbolism should be understood. We have described, term for term, the numbers in equation (I-40) by their thermodynamic symbols in equation (I-41) to enable the student to make a one-to-one correspondence between the numbers and the symbols. The symbol $\overline{H}^\circ_{1550^\circ\mathrm{C}}(\mathrm{CO})$, for example, stands for the enthalpy (H) of one mole (the bar over the H) of carbon monoxide (the CO in parenthesis) in its standard state (the superscript zero) at 1550°C (the subscript 1550°C).

In order to complete our discussion, we must discuss the method by which the quantities $\overline{H}_{\mathrm{T}_2}{}^\circ - \overline{H}_{\mathrm{T}_1}{}^\circ$ are evaluated. If we substitute ΔH for q_p [see equation (I-33)] in equation (I-19) we obtain

$$C_p = \lim_{\Delta\mathrm{T}\to 0} \Delta H/\Delta\mathrm{T})_p \tag{I-42}$$

or

$$C_p = \left.\frac{\partial H}{\partial \mathrm{T}}\right)_p \tag{I-43}$$

That is, the heat capacity at constant pressure is equal to the derivative of the enthalpy with respect to temperature at constant pressure. Since the standard states are, by definition, at a constant pressure of one atmosphere, we need not specify "constant pressure" in the equation

$$C_p{}^\circ = \frac{dH^\circ}{d\mathrm{T}} \tag{I-44}$$

Thus, if we know $C_p{}^\circ$ as a function of temperature (such quantities are tabulated in standard compilations, such as those listed in Appendix I), then we can obtain

$$H_{\mathrm{T}_2}{}^\circ - H_{\mathrm{T}_1}{}^\circ = \int_{\mathrm{T}_1}^{\mathrm{T}_2} C_p{}^\circ\, d\mathrm{T} \tag{I-45}$$

In fact we have made just such a calculation for the mixture of $\frac{1}{2}$ mole of N_2(g) and $\frac{3}{2}$ mole of H_2(g) with $T_2 = 500°C$ on p. 14.

If there is a phase transition, for example melting, between T_1 and T_2 we have

$$H_{T_2}{}^\circ - H_{T_1}{}^\circ = \int_{T_1}^{T_t} C_p{}^\circ \, dT + \Delta H_t + \int_{T_t}^{T_2} C_p{}^\circ dT \qquad \textbf{(I-46)}$$

where the integration from T_1 to the transition temperature, T_t, is over the heat capacity of the form stable between T_1 and T_t, ΔH_t is the enthalpy change accompanying the transition, and the integration from T_t to T_2 is over the heat capacity of the form stable between T_t and T_2.

If we now consider the chemical reaction of equation (I-37) and the corresponding equation (I-41) relating the standard enthalpy change at 1550°C to the standard enthalpies of formation at 25°C, we note that we can substitute according to equation (I-45) and obtain

$$\begin{aligned} \Delta H^\circ_{1550°C} &= 3\,\Delta H^\circ_{f,25°C}(CO) - \Delta H^\circ_{f,25°C}(Fe_2O_3) \\ &+ 3\int_{25°C}^{1550°C} C_p{}^\circ(CO)\,dT + 2\int_{25°C}^{1536°C} C_p{}^\circ(Fe)\,dT + 2\,\Delta H_{\text{fusion}}(Fe) + 2\int_{1536°C}^{1550°C} C_p(Fe)\,dT \\ &- \int_{25°C}^{1550°C} C_p{}^\circ(Fe_2O_3)\,dT - 3\int_{25°C}^{1550°C} C_p{}^\circ(C)\,dT \end{aligned} \qquad \textbf{(I-47)}$$

which we can write

$$\Delta H^\circ_{1550°C} = \Delta H^\circ_{25°C} + \int_{25°C}^{1550°C} \Delta C_p{}^\circ \, dT + \Delta H_{\text{fusion}}\,(Fe) \qquad \textbf{(I-48)}$$

where

$$\Delta C_p{}^\circ = 3\bar{C}_p{}^\circ(CO) + 2\bar{C}_p{}^\circ(Fe) - \bar{C}_p{}^\circ(Fe_2O_3) - 3\bar{C}_p{}^\circ(C)$$

For the general reaction

$$aA + bB + \cdots = lL + mM + \cdots \qquad \textbf{(I-49)}$$

and any two temperatures we similarly obtain

$$\Delta H_{T_2}{}^\circ = \Delta H_{T_1}{}^\circ + \int_{T_1}^{T_2} \Delta C_p{}^\circ \, dT \qquad \textbf{(I-50)}$$

where

$$\Delta C_p{}^\circ = l\bar{C}_p{}^\circ(L) + m\bar{C}_p{}^\circ(M) - a\bar{C}_p{}^\circ(A) - b\bar{C}_p{}^\circ(B) + \cdots \qquad \textbf{(I-51)}$$

and we have assumed that none of the condensed phases involved have undergone phase transitions.

SUMMARY

In this chapter we have considered the concepts energy, heat, and work, and have inferred the existence of the property internal energy from an adiabatic work experiment. We then defined heat as the difference between the internal energy change and the work done on the system. We have discussed why heat added at constant volume and heat added at constant pressure are changes in properties, and utilized this fact to determine standard heats of any reaction at constant P or V from a knowledge of the appropriate standard heats of formation of the products and the reactants. We have then discussed how the heat of a reaction at any temperature can be calculated from the heats of formation of the products and of the reactants at a given temperature, usually 25°C and from a knowledge of the heat capacities of the products of the reactants.

In summary, the formulae which we have developed are

$$\Delta U = q + w = q - \int_{V_1}^{V_2} P_e \, dV\text{: closed system; only } P\text{–}V \text{ work.}$$

$$q_p = \Delta H.$$

$$q_v = \Delta U.$$

$$C_p = \lim_{\Delta T \to 0} \left.\frac{\Delta H}{\Delta T}\right)_P = \lim_{\Delta T \to 0} \frac{q_p}{\Delta T}\text{, and similarly for } C_v \text{ and } \Delta U.$$

$$q_p = C_p \, dT.$$

For a reaction at 298°K,

$$\Delta H = \sum_{\text{prod.}} \nu_i H_{i,f}^{\circ} - \sum_{\text{reactants}} \nu_i H_{i,f}^{\circ}$$

where ν_i are stoichiometry numbers.

$$\Delta H_T^{\circ} = \Delta H_{298}^{\circ} + \int_{298}^{T} \Delta C_P^{\circ} \, dT$$

Problems

The problems below are divided into two sections. The problems in Section A are physical in nature while those in Section B are chemical in nature.

Section A

I-1 A gas in an adiabatically insulated container expands against a piston so as to raise a 1000 gram mass a height of 10 cm. The gas is then placed in contact with a heat source and 0.10 calorie is added to the gas. How much work must now be done to adiabatically return the gas to its initial state? The following constants and conversion factors are useful in working this problem: $g = 980\ \text{cm/sec}^2$; $1\ \text{cal} = 4.184 \times 10^7$ ergs.

I-2 A 10 gram lead bullet is fired vertically *into* a 990 gram lead block which is hanging on a string. The block and bullet rise to a height of 20.4 cm before dropping back to the end of the string. Calculate the temperature rise of the block and bullet assuming that they are initially at the same temperature and that they lose no heat. Neglect the string in all calculations. *Hint:* Use the Law of Conservation of Momentum to obtain the initial velocity of the block. The following constants and conversion factors are useful in working the problem:

$$C_p(\text{Pb}) = 0.031\ \text{cal/deg-gram};\ g = 980\ \text{cm/sec}^2;\ 1\ \text{cal} = 4.184 \times 10^7\ \text{ergs}.$$

I-3 The power incident on the surface of the earth from the sun is about 1.2 kilowatts/m^2 on a clear day. Calculate the rate at which water vaporizes from 1 m^2 of ocean surface if the temperature is constant and the only means of cooling the water is vaporization. Take the heat of vaporization of water to be 539 cal/gram. The following conversion factors are useful: 1 watt = 1 joule/sec; 1 joule = 0.239 cal.

I-4 One hundred grams of water at 50°C are mixed with 10 grams of ice at 0°C in an adiabatically insulated system. Calculate the final temperature of the system. Take C_p for H_2O to be 1 (cal/deg-gram) and $\Delta H^\circ_{\text{fusion}} = 80$ cal/gram.

I-5 An engine is a device which converts heat into work. As we shall see in the next chapter there are limitations on the conversion, namely the heat taken in is not converted entirely into work, but some heat is exhausted. Consider an engine which takes some substance through a *cycle*. In so doing it takes in q_2 calories and exhausts q_1 calories. How much work is done in a cycle? The heat for which you "pay" is q_2. How would you measure the efficiency of such an engine? If $q_2 = 1$ kilocalorie and $q_1 = \frac{1}{2}$ kilocalorie, how high could the engine lift 1 kg in one cycle? What is the efficiency of the engine? The following constants and conversion factors are useful in working this problem:

$$g = 980\ \text{cm/sec}^2;\ 1\ \text{cal} = 4.184 \times 10^7\ \text{ergs}.$$

I-6 A gas is expanded from 1 liter to 19 liters against a pressure which is given by

$$P_{\text{ex}} = \frac{1}{1 + V}\ \text{atm}$$

During this process 298 calories of heat are added to the system. What is w, the work done on the system? How much work must be done on the system to return it to its original state adiabatically? Conversion factor:

$$1\ \text{liter-atm} = 24.2\ \text{cal}$$

I-7 Calculate the work done on one mole of an ideal gas to compress it isothermally and reversibly at 298°K, from 20 liters to 2 liters. Recall that a reversible expansion implies that the external pressure is equal to the internal pressure which is given by the equation of state.

Constant: R = 1.987 cal/mole deg

Section B

I-8 The enthalpies of formation of $Al_2O_3(s)$ and $Al_2O(g)$ at 298°K are −400.4 kcal and −31.4 kcal, respectively. Calculate the enthalpy change at 298°K for the reaction

$$Al_2O_3(s) = Al_2O(g) + O_2(g)$$

I-9 The enthalpies of formation of lithium dimer ($Li_2(g)$) and lithium monomer (Li(g)) at 0°K are 50.30 kcal and 38.03 kcal, respectively. What is the dissociation energy of the Li—Li bond?

I-10 The standard enthalpies of formation of boron trihydride ($BH_3(g)$) and diborane ($B_2H_6(g)$) at 0°K are 26.41 kcal and 13.55 kcal, respectively. What is the dissociation energy of diborane into boron trihydride?

I-11 The standard heats of formation of $SF_6(g)$, S(g) and F(g) at 0°K are −288.5 kcal, 66.1 kcal and 18.4 kcal, respectively. What is the average S—F bond energy in $SF_6(g)$? The heat of formation of $SF_4(g)$ at 0°K is −172.2 kcal. What is the average S—F bond energy in $SF_4(g)$?

I-12 The standard heats of formation of $H_2O_2(g)$, OH(g) and O(g) at 0°K are −31.03 kcal, 9.29 kcal and 58.99 kcal, respectively. What are the standard enthalpy changes for the following reactions?

$$H_2O_2(g) = 2\,OH(g)$$

and

$$O_2(g) = 2\,O(g)$$

How do you account for the difference between the energies of the O—O bonds in $H_2O_2(g)$ and $O_2(g)$?

I-13 The heats of formation of $C_2H_2O_4(s)$ (oxalic acid), $H_2O(l)$ and $CO_2(g)$ at 298°K are −197.6 kcal, −57.8 kcal and −94.05 kcal respectively. Calculate the heat of combustion of $C_2H_2O_4(s)$ at 298°K.

I-14 The enthalpies of formation of NaCl(g), Na(g), Cl(g), $Na^+(g)$ and $Cl^-(g)$ are −43.50 kcal, 25.98 kcal, 29.01 kcal, 146.01 kcal and −58.3 kcal, respectively. Is it energetically more favorable for NaCl(g) to dissociate according to

$$NaCl(g) = Na^+(g) + Cl^-(g)$$

or

$$NaCl(g) = Na(g) + Cl(g)?$$

Explain why it might be otherwise in solution.

I-15 An important reaction in the production of pig iron from iron ore is

$$CaCO_3(s) + SiO_2(l) + C(s) = CaSiO_3(l) + 2CO(g)$$

Given is the following information:

	$CaCO_3$(s)	SiO_2(s)	C(s)	$CaSiO_3$(s)	CO(g)
ΔH_f° 25°C $\frac{\text{Kcal}}{\text{mole}}$	−288.45	−209.6	0	−382.9	−26.417
$\overline{C}_p$ cal/mole deg	30	15	4	30	7

The heat capacity of $CaSiO_3$(l) is 30 cal/mole deg, the melting point is about 1450°K, and the heat of fusion is 3 cal/mole. The heat capacity of SiO_2(l) is 16 cal/mole deg, the melting point is 800°K, and the heat of fusion is 2 kcal/mole. Calculate ΔH°_{1550} for the above reaction assuming constant heat capacities in the solid and liquid ranges.

CHAPTER II

The Second Law, A Criterion for Reversibility

At this point, it might pay to pause and consider philosophically the position in which we find ourselves regarding the relationships between energy, heat, and work. We have said simply that a change in energy must be accounted for in terms of "heat" type energy and "work" type energy. It is the *sum* of these contributions that is a change in a property, namely **internal energy,** U. A rather loose, commonly used statement of the first law is "you can't get something for nothing, you can only break even." We are now in a position to ask "under what circumstances can we break even?" The answer will lead to the definition of absolute temperature, and hopefully, to a better understanding of the concept "temperature." It will also lead to the definition of a new property which is of great utility in helping chemists decide whether or not certain reactions will proceed under specified conditions. This property is given the name **entropy** from the Greek *entropia*, meaning "transformation."

Let us, before stating the second law, think of an experiment the results of which are experientially obvious. Consider two isolated blocks of copper having the same mass, say 63.54 g, one being at 100°C and the other at 0°C. We place the blocks in thermal contact with each other and ask, "What happens?" A possible answer, from the point of view of the first law, is that the block initially at 100° heats to 105°, and the block initially at 0° cools to −5°C. This is to say, the amount of heat $n\overline{C}_p(105 - 100)$ has passed into the hotter block, and the equal amount $n\overline{C}_p\ (5 - 0)$ has passed *from* the cooler block. This process certainly is *possible*, and does not violate the first law; after all, $w = 0$ and energy (in this case in the form of heat) is conserved.* The plain facts of the matter are that none of us would be willing to hold his breath until such a process occurred, even though it is not prevented from doing so by the first law! Clearly the first law is not the last word regarding energy interchanges. The first law, in fact, says *nothing* about spontaneity, and it is just this question in which chemists have the most interest. Not only do we want to determine what would be the energy interchange in a reaction, but we would also like to predict the direction in which a reaction will go. For example, consider the reaction

$$N_2(g) + 1/2\, O_2(g) = N_2O(g)$$

Realizing full well that the air that we breath consists of roughly 80% nitrogen and 20% oxygen, we all have a vested interest in having this reaction not occur, and we would be very uneasy if some individual invented a catalyst which would enable the reaction to proceed spontaneously to completion, thus converting our atmosphere to laughing gas. While the situation might be hilarious for a short time, we would, in fact, die laughing. It is obvious that N_2O is not nearly as useful to the breathing process as is a solution of N_2 and O_2. Have no fear! We have been able, by making use of second law calculations, to assure ourselves that, under present atmospheric conditions, such a conversion is impossible.

THE SECOND LAW OF THERMODYNAMICS

The second law, then, along with energy interchange, concerns man's experience with spontaneity and equilibrium. There are several statements of the second law, all of which may be shown to be equivalent. One statement, correlating well with our copper block experiment, is as follows:

It is impossible for heat to flow from a cold body to a hotter body with no changes taking place in the surroundings; i.e., heat does not spontaneously flow from a lower to a higher temperature.

---

* We have neglected the small volume changes with temperature change here, and thus the *PV* work which is negligible compared to the quantity of energy about which we are talking.

This statement is due to Clausius (1822–1888) who was a mathematical physicist at the University of Zurich. The following statement of the law is that of William Thomson, better known as Lord Kelvin, a professor of natural philosophy at Glasgow from 1846 to 1889.

It is impossible by a cyclic process to convert heat from a reservoir into work without, in the same operation, passing some heat into a lower temperature reservoir.

Basically the above statements embody one simple idea, namely that processes which occur spontaneously in isolation, have a "natural" direction. The natural direction for heat flow is from a hot to a cold body, the natural tendency is for a gas to expand into a vacuum, the natural tendency is for ice to melt at temperatures above its melting point, and, as we shall show, the natural direction for chemical reactions in isolation is in the direction of increasing entropy. Thus a good deal of our discussion in this chapter will be directed towards defining that most useful property S, the **entropy.** While there are a number of ways of introducing the property called entropy ranging from pure definition to rather abstract mathematical formalism, the traditional approach *via* discussion of heat engines is perhaps the most appropriate here. The heat engine approach has the advantage of providing a direct physical handle on the entropy concept. It has the slight disadvantage of requiring the student to become familiar with an abstract cyclic process not obviously related to chemistry, namely the **Carnot Cycle.** It is this process which we now discuss.

THE CARNOT CYCLE

Consider a system being carried through the following cyclic process in four steps as illustrated in Fig. 2.1.*

Step I: Isothermal expansion along Path I at temperature t_2.
Step II: Adiabatic expansion along Path II. $q = 0$.

---

* We note that this is not the "usual" appearance of the Carnot cycle on a $P-V$ diagram, but we have tired of the "usual" appearance, and the student will also soon come into tiring contact with the usual appearance. This diagram is one which is observed for some working substances and has the additional advantage of not tying the student down to thinking in terms of an ideal gas as a working substance.

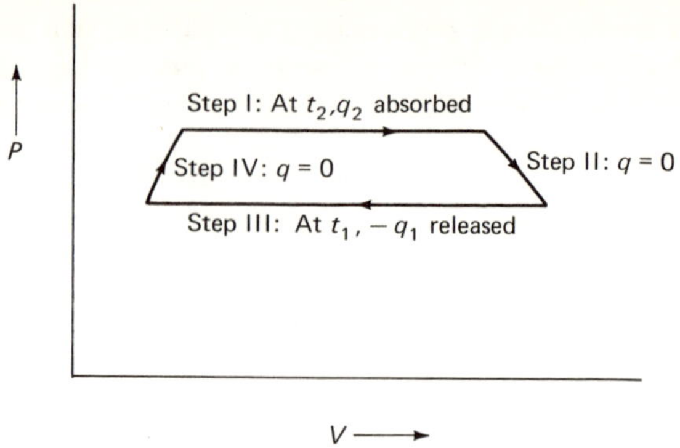

Figure 2.1 Steps in a Carnot Cycle.

Step III. Isothermal compression along Path III at temperature $t_1 < t_2$.

Step IV: Adiabatic compression along Path IV. $q = 0$.

In this case we use lower case t to indicate specifically that these temperatures are *empirical temperatures.* We note that in this process any heat transfer occurs at constant temperature and any temperature change comes about adiabatically. If all changes occur reversibly, the above process is called a **reversible Carnot cycle,** and it is this which we wish to consider. The second law applied to such a cycle leads directly to the definition of thermodynamic temperature in terms of the efficiency of a reversible Carnot engine. The efficiency is defined as the ratio of the work done *by* the system, $-w$, to the heat, q_2, added at the higher temperature t_2, i.e., the heat for which you pay in a real case. The first law allows for an efficiency of unity corresponding to a cyclic process in which all of the heat bought is converted into work. An efficiency of unity would violate the Kelvin statement of the second law and would therefore be out of accord with the totality of human experience, for in this case heat would be quantitatively converted into work in a cycle. If such were possible, for example, an automobile could be run on the thermal energy from any source, such as the thermal energy in the air.

Applying the first law to one cycle of the Carnot engine we have

$$\Delta U = 0 = q + w \qquad \textbf{(II-1)}$$

$$= q_2 + q_1 + w \qquad \textbf{(II-2)}$$

for it is clear that ΔU for any cyclic process is zero. Thus

$$-w = q_2 + q_1 \tag{II-3}$$

and, hence, the efficiency is given by

$$e = \frac{\text{work done}}{\text{heat ``bought''}} = -\frac{w}{q_2} = \frac{q_2 + q_1}{q_2} \tag{II-4}$$

We shall now demonstrate that *no* Carnot engine is more efficient than a reversible Carnot engine by showing that if such were the case the Clausius statement would be violated.

Consider two Carnot engines, A (not necessarily reversible) and B (reversible) linked together such that both operate between t_2 and t_1 ($<t_2$). Let the work w_A obtained from A running "forward" drive B, with B running *reversibly* as a heat "pump" (this is always possible since B is a reversible engine). We illustrate this in Fig. 2.2.

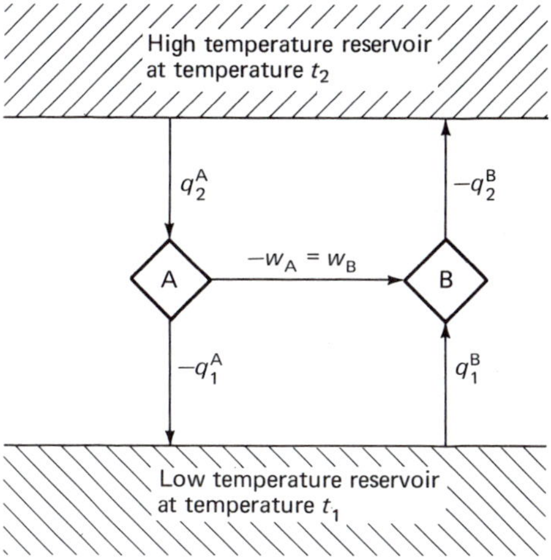

Figure 2.2 Coupling of two Carnot engines to consider maximum efficiency possible.

Suppose that B, the reversible engine, were less efficient than A. This would mean that *if* B and A were each operating in the forward direction and doing the same work then $q_2^A < q_2^B$. This would also mean that with B operating in reverse, and no net work by the combined engines, the net effect would be that A takes from the reservoir at t_2 less heat than B puts into it, i.e., heat is pumped from a low to a high temperature reservoir with no work being done. We demonstrate this mathematically as follows. Since we set the work done by A equal to the work done on B

$$q_2^{\mathrm{A}} + q_1^{\mathrm{A}} = -w_{\mathrm{A}} = w_{\mathrm{B}} = -(q_1^{\mathrm{B}} + q_2^{\mathrm{B}}) \tag{II-5}$$

and thus

$$q_2^{\mathrm{B}} + q_1^{\mathrm{B}} = -(q_2^{\mathrm{A}} + q_1^{\mathrm{A}}) \tag{II-6}$$

This equation indicates that the only net changes which occur when A operates one cycle in a forward direction and B operates one cycle in a reverse direction are heat transfers from one temperature heat source (or sink) to another. But in which direction is the heat transferred in the total cycle? To answer this question we examine the supposed relationship between the efficiencies of A and B

$$e_{\mathrm{A}} > e_{\mathrm{B}} \tag{II-7}$$

that is

$$\frac{q_2^{\mathrm{A}} + q_1^{\mathrm{A}}}{q_2^{\mathrm{A}}} > \frac{q_2^{\mathrm{B}} + q_1^{\mathrm{B}}}{q_2^{\mathrm{B}}} \tag{II-8}$$

But if we divide both sides of equation (II-6) by q_2^{B} we obtain

$$\frac{q_2^{\mathrm{B}} + q_1^{\mathrm{B}}}{q_2^{\mathrm{B}}} = -\left(\frac{q_2^{\mathrm{A}} + q_1^{\mathrm{A}}}{q_2^{\mathrm{B}}}\right) \tag{II-9}$$

We can substitute the latter part of this expression for the right-hand side of equation (II-8) and obtain

$$\frac{q_2^{\mathrm{A}} + q_1^{\mathrm{A}}}{q_2^{\mathrm{A}}} > -\left(\frac{q_2^{\mathrm{A}} + q_1^{\mathrm{A}}}{q_2^{\mathrm{B}}}\right) \tag{II-10}$$

Since the numerators of both sides of this inequality are the same, and since both sides are positive ($q_2^{\mathrm{A}} + q_1^{\mathrm{A}} > 0$, $q_2^{\mathrm{A}} > 0$ and $q_2^{\mathrm{B}} < 0$), it follows that*

$$q_2^{\mathrm{A}} < -q_2^{\mathrm{B}} \tag{II-11}$$

But this means exactly that the amount of heat withdrawn from the heat source at t_2 (the higher temperature) by engine A is less than the amount of heat transferred to the same "heat source" by the reversible Carnot heat pump, and thus that there is a net heat transfer to this reservoir. Thus, since by equation (II-6) the source of this

* Consider the analogous inequalities

$$\frac{1}{2} > -\frac{1}{(-3)} \quad \text{and} \quad 2 < -(-3)$$

heat is the reservoir at t_1, we have demonstrated that if a Carnot engine more efficient than a reversible Carnot engine exists, then it is possible to accomplish a net transfer of heat from a body at a lower temperature (t_1) to a body at a higher temperature (t_2) with no change in the surroundings (the engines went through a cycle). But this is a contradiction of the Clausius statement of the second law, and thus, if the second law is valid, our hypothesis is incorrect and there is no Carnot engine which is more efficient than a reversible one. It is a corollary of this statement that all reversible Carnot engines operating between the same two temperatures are equal in efficiency, for clearly if one reversible Carnot engine were more efficient than another, there would be an engine more efficient than a reversible Carnot engine.

The argument to this point has, for those interested in chemistry, perhaps lost some charm because we appear to be very far afield from chemistry. Unfortunately, things are going to get worse before they get better. In order to utilize the result, seemingly unrelated to chemistry but in truth extremely powerful, that no Carnot engine operating between two given temperatures is more efficient than a reversible Carnot engine operating between the same two temperatures, we must make a somewhat abstract argument. The level of this argument is, however, one that should be followed easily if attention is paid to details. The result is that we define the absolute, or thermodynamic, temperature scale as that scale in which a reversible Carnot engine has unit efficiency if its sink temperature, t_1, is zero. The scale is called the **Kelvin Scale** when the size of the degree is such that 273.16°K (Kelvin) is the absolute temperature at which water vapor, solid water, and liquid water coexist in equilibrium. This temperature is called the **triple point temperature** of water.

The argument defining this scale in terms of the efficiency of a reversible Carnot engine involves a bit of algebra, so before examining the "trees" in detail, it might just pay to consider an outline of the "forest." The argument is essentially outlined as follows:

(a) We note that since all reversible Carnot engines operating between the same source and sink temperatures have the same efficiencies, the efficiency of a reversible Carnot engine must depend only upon the source and sink temperatures.

(b) We envision two reversible Carnot engines coupled as shown in Fig. 2.3a, such that one (engine A) operates from a source temperature t_3 to a sink temperature t_2, and the other (engine B) utilizes the

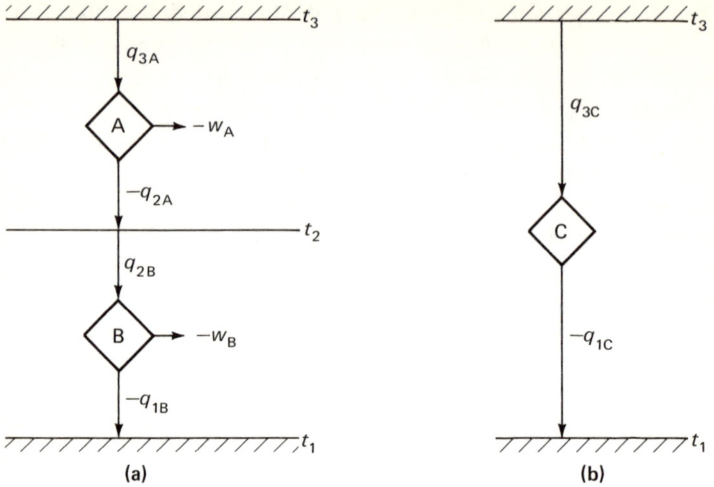

Figure 2.3 (**a**) The equivalence of two Carnot engines, A and B, operating between sources and sinks t_3, t_2 and t_2, t_1 respectively, to (**b**) one Carnot engine, C, operating between source and sink t_3, t_1. $q_{2B} = -q_{2A}$ and $w_C = w_A + w_B$. This equivalence will be further discussed *via* consideration of Carnot Cycles (see p. 38).

heat exhausted from A as a source at t_2, and exhausts at its sink, t_1. The coupled system A and B is equivalent to a single engine C operating between t_3 and t_1, as illustrated in Fig. 2.3b.

(**c**) We note that the efficiencies of A and of B are dependent on t_3 and t_2, and on t_2 and t_1 respectively, and the efficiency of C is dependent only on t_3 and t_1.

(**d**) We conclude that the results of observations (b) and (c) imply a definite functional form for the efficiency of Carnot engines relating to their source and sink temperatures. This functional form is used to define the absolute temperature scale.

We now investigate the "trees" in detail, starting with point (a) above. The second law demands that all reversible Carnot engines operating between the same source and sink temperatures have the same efficiency, and hence that the efficiencies of reversible Carnot engines are functions of the source and sink temperatures only. We symbolize this dependence of the efficiency of a reversible Carnot engine, with its associated reversible Carnot Cycle, as illustrated in Figures 2.4a and 2.4b, as follows:

$$e = F(t_1, t_2) \quad \text{(II-12)}$$

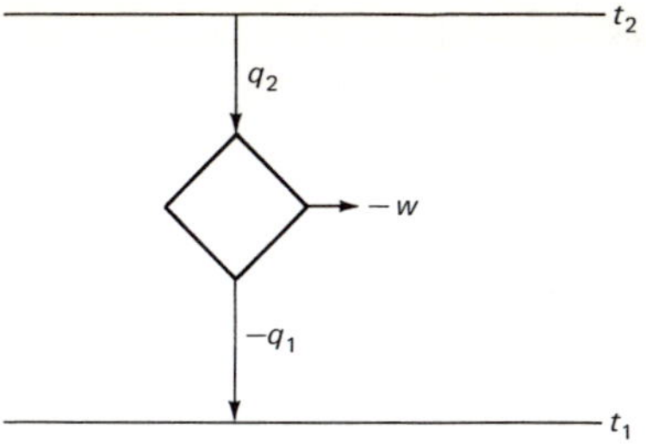

(a)

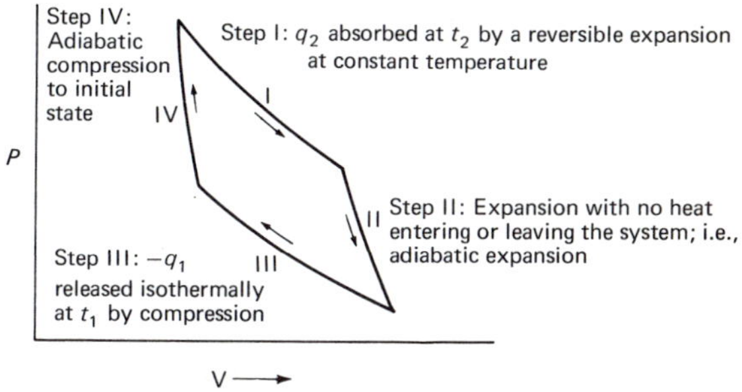

(b)

Figure 2.4 **(a)** Reversible Carnot engine operating between source t_2 and sink t_1. q_2 is absorbed at t_2, $-q_1$ is released at t_1, and $-w$ is done by the engine. **(b)** *P–V* representation of the Carnot Cycle carried out by the engine represented in Fig. 2.4a. The difference, between this *P–V* representation and that shown in Fig. 2.1 is the working substance used to drive the engine.

We note that since

$$e = (q_2 + q_1)/q_2 = 1 + q_1/q_2 \qquad \textbf{(II-13)}$$

and since e, and therefore $e - 1$, depends upon t_1 and t_2, the ratio q_1/q_2 in general depends only upon t_1 and t_2:

$$q_1/q_2 = f(t_1, t_2) \qquad \textbf{(II-14)}$$

If we now consider two reversible Carnot engines, A and B, operating between t_3 and t_2, and between t_2 and t_1 respectively, as diagrammed in Fig. 2.3a, coupled in such a manner that the heat exhausted from A at t_2 is absorbed in B at t_2, we have

$$e_A = (q_{3A} + q_{2A})/q_{3A} = 1 + q_{2A}/q_{3A} \qquad \textbf{(II-15)}$$

or, as with equation (II-13),

$$q_{2A}/q_{3A} = f(t_2, t_3) \qquad \textbf{(II-16)}$$

Similarly, for engine B,

$$q_{1B}/q_{2B} = f(t_1, t_2) \qquad \textbf{(II-17)}$$

where the same function, f, *is applicable to both cycles because they are reversible Carnot Cycles!*

We next observe that when the two Carnot engines, A and B, traverse one cycle, the net effect is equivalent to one Carnot engine, C, operating between t_3 and t_1. For engine C, therefore, we may write

$$q_{1C}/q_{3C} = f(t_1, t_3) \qquad \textbf{(II-18)}$$

where f is again the function appropriate to all reversible Carnot engines.

It is clear from the choice of Carnot engines made above that

$$q_{1C} = q_{1B} \qquad \textbf{(II-19)}$$

$$q_{2A} = -q_{2B} \qquad \textbf{(II-20)}$$

and

$$q_{3C} = q_{3A} \qquad \textbf{(II-21)}$$

Combining equations (II-16) and (II-21), we obtain

$$q_{2A}/q_{3C} = f(t_2, t_3) \qquad \textbf{(II-22)}$$

Similarly, combining equations (II-17) and (II-19), we obtain

$$q_{1C}/q_{2B} = f(t_1, t_2) \qquad \textbf{(II-23)}$$

Multiplying (II-22) by (II-23) yields

$$(q_{1C}/q_{2B}) \cdot (q_{2A}/q_{3C}) = f(t_1, t_2) \cdot f(t_2, t_3) \qquad \textbf{(II-24)}$$

But since $q_{2A} = -q_{2B}$, then

$$-q_{1C}/q_{3C} = f(t_1, t_2) \cdot f(t_2, t_3) \qquad \textbf{(II-25)}$$

Comparing equation (II-18) with equation (II-25), we immediately see that

$$-f(t_1, t_3) = f(t_1, t_2) \cdot f(t_2, t_3) \qquad \textbf{(II-26)}$$

At this point, it is necessary to examine the effect of interchanging t_2 and t_3 in $f(t_2, t_3)$. Recalling the definition of $f(t_2, t_3)$:

$$e_A - 1 = f(t_2, t_3)$$

Now, to suit our private ends, we inquire into the effect of operating engine A in reverse, i.e., as a heat *pump* (an example of which is a refrigerator). We observe that the *magnitudes* of the heats and the work remain the same, but the *signs* of all quantities reverse, e.g., q_{2A} is absorbed at t_2, and $-q_{3A}$ is exhausted at the higher temperature, t_3. Thus, for engine A operating as a heat pump,

$$e_A = (q_{3A} + q_{2A})/q_{2A} \tag{II-27}$$

or

$$e_A - 1 = q_{3A}/q_{2A} = f(t_3, t_2) \tag{II-28}$$

By comparison with equation (II-16), we see that

$$f(t_2, t_3) = 1/f(t_3, t_2) \tag{II-29}$$

Therefore the effect of interchanging temperatures is to invert the function. Making use of this fact, we can arrange equation (II-25) to read

$$-f(t_1, t_3) = f(t_1, t_2)/f(t_3, t_2) \tag{II-30}$$

Since the left side of equation (II-30) does *not* depend upon t_2, neither must the right side. Therefore

$$-f(t_1, t_3) = f(t_1, t_2)/f(t_3, t_2) = F(t_1)/F(t_3) \tag{II-31}$$

By comparison with equations (II-30) and (II-18),

$$-q_{1C}/q_{3C} = F(t_1)/F(t_3) \tag{II-32}$$

We *define* the thermodynamic temperature scale by this relation! The *thermodynamic temperatures* T_1 and T_3 corresponding to empirical temperatures t_1 and t_3 are $T_1 = F(t_1)$ and $T_3 = F(t_3)$.

ABSOLUTE TEMPERATURE

Substituting the defined thermodynamic temperatures into equation (II-32) yields

$$-q_1/q_3 = T_1/T_3 \tag{II-33}$$

where the subscript C has been dropped, since equation (II-33) applies to *all* reversible Carnot engines operating between T_1 and T_3. In terms of the efficiency of any reversible Carnot engine operating between source T_3 and sink T_1,

$$e = 1 + q_1/q_3 = 1 - T_1/T_3 \tag{II-34}$$

We see immediately from this equation that the efficiency of a reversible Carnot engine can be equal to unity (all heat converted into work) only if the sink temperature, T_1, is zero. Thus we answer the question asked at the beginning of the chapter. The answer to "under what conditions can we break even?" is "at a thermodynamic sink temperature of zero." Note particularly that if a negative thermodynamic temperature were attainable, it would be possible to construct a Carnot engine with an efficiency *greater* than unity. This engine would clearly violate the First Law, so to the extent that the first and second laws are valid, 0° absolute is *absolutely* the lowest temperature attainable!

Having now answered the question, "Where can we break even?" we are in a position to pursue further the train of thought to its most important conclusion, namely, the definition of the entropy function, and criteria of spontaneity in chemical systems in terms of the entropy.

ENTROPY

It does not take much playing with the relationships which one finds from a consideration of reversible Carnot cycles to realize that for a number of given cycles

$$\sum_{\substack{i \\ \text{cycle}}} \frac{q_i^{\text{rev.}}}{T_i} = 0$$

Thus, for a reversible Carnot engine operating between thermodynamic temperatures T_1 and T_2, we have seen that

$$-\frac{q_1^{\text{rev.}}}{q_2^{\text{rev.}}} = \frac{T_1}{T_2} \quad \text{or} \quad \frac{q_1^{\text{rev.}}}{T_1} + \frac{q_2^{\text{rev.}}}{T_2} = 0 \qquad \text{(II-35)}$$

Now the fact that an integral is defined as an infinite sum, the fact that for a *property* an integral around a closed path, i.e., over a cycle, is zero, and the fact that we are vitally interested in finding new and useful properties, all lead us to look very closely at any sum which is zero over a closed path. We therefore want to look very carefully at the fact that for a collection of reversible Carnot Cycles

$$\sum_{\substack{i \\ \text{cycle}}} \frac{q_i^{\text{rev.}}}{T_i} = 0 \qquad \text{(II-36)}$$

We will argue, in fact, that any closed *reversible* path may be approximated to any degree of accuracy as an infinite sum of infinitesimal reversible Carnot Cycles with heats dq_i being absorbed at temperatures

T_i, and so

$$\sum_{\substack{i \\ \text{any cycle}}} \frac{dq_i^{\text{rev.}}}{T_i} = 0 \tag{II-37}$$

and, in the limit

$$\int_{\substack{\text{any} \\ \text{cycle}}} \frac{dq^{\text{rev.}}}{T} = 0 \tag{II-38}$$

We argue this in detail in the following manner: Consider, for example, a single gaseous substance being carried through a Carnot cycle as exhibited on a *P–V* diagram (Fig. 2.5a). We might equally well

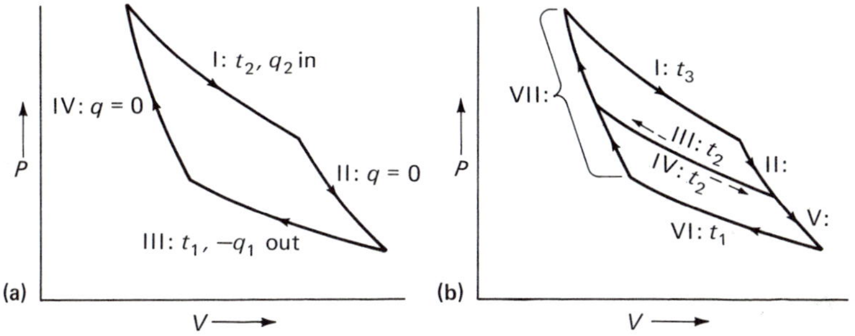

Figure 2.5 Illustration of one Carnot Cycle (left) being equivalent to a sum of two Carnot Cycles (right).

traverse the four-step path I, II, III, IV in seven steps, as shown in Fig. 2.5b, thus effectively replacing one cycle by two. Since each step is reversible, however, it is clear that at the intermediate isotherm t_2, the work and heat in Step III are just the negative of the work and heat in step IV, so that Steps III and IV taken together exactly cancel. In fact, *any* steps taken first in one and then in a reverse direction along a line common to two cycles cancel, as for the two cycles shown in Fig. 2.5b. Thus the two cycles in Fig. 2.5b are equivalent to the cycle of Fig. 2.5a. Now consider an arbitrary reversible cycle shown in Fig. 2.6a, where the temperature is varying throughout the process. Let us place this figure on a net of adiabats and isotherms,* as in Fig. 2.6b, and roughly traverse the arbitrary path on reversible isotherms and adiabats as shown in the bold-faced outline. It is clear

* An adiabat is a line along which $q = 0$. An isotherm is a line along which t is constant.

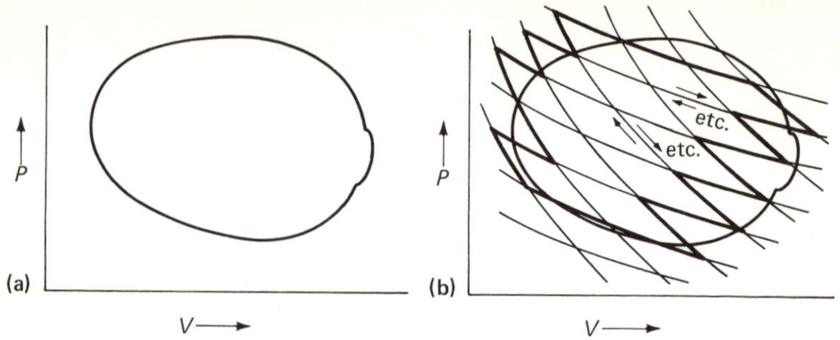

Figure 2.6 (**a**) Arbitrary cycle. (**b**) Approximation of an arbitrary cycle on a net of adiabats and isotherms.

that, as before, the bold outline is equivalent to all of the reversible Carnot cycles which it encloses and that the sum of these cycles *is* equivalent to the cycle described by the outline. For all of the reversible Carnot cycles within the outline

$$\sum \frac{q_i^{\text{rev.}}}{T_i} = 0 \tag{II-39}$$

Using the by now familiar process of approximating a smooth curve by a large number of small finite steps, (in this case adiabats and isotherms) and passing from approximation to exactitude by taking the limit, we have,

$$\lim_{\text{max.}\, q_i \to 0} \sum_{\substack{\text{reversible} \\ \text{cycles}}} \frac{q_i^{\text{rev.}}}{T_i} = \int_{\text{cycle}} \frac{dq^{\text{rev.}}}{T} = 0 \tag{II-40}$$

This is to say, we make the "net size" of adiabats and isotherms smaller and smaller until in the limit the arbitrary reversible cycle is traversed exactly by reversible adiabats and isotherms, i.e., is traversed by reversible Carnot cycles. Thus we have proved that around an arbitrary *reversible* cycle

$$\int_{\text{cycle}} \frac{dq^{\text{rev.}}}{T} = 0 \tag{II-41}$$

What does this mean physically? It means that no matter what reversible cycle a system traverses, the integral of $dq^{\text{rev.}}/T$ around the cycle is zero. This is not true of $dq^{\text{rev.}}$, for $q^{\text{rev.}}$ is not a property. That it is true for $dq^{\text{rev.}}/T$ means that there exists a property, which is conventionally given the symbol S and called entropy, the change in which can be evaluated for a given process according to

$$S_2 - S_1 = \int_1^2 \frac{dq^{\text{rev.}}}{\text{T}} \tag{II-42}$$

or

$$dS = dq^{\text{rev.}}/\text{T} \tag{II-43}$$

Hence we can determine the difference between the entropy in two different states by integrating $dq^{\text{rev.}}/\text{T}$ over *any* reversible path connecting the states. A question which might be asked is "why can we not similarly approximate any *irreversible* cycle by a series of *reversible* Carnot cycles and thus prove that $q^{\text{irrev.}}/\text{T}$ is a property?" The answer is that the states of the working substance of the engine will necessarily be different during reversible and irreversible cycles, and thus the irreversible cycle can never be equivalent to the limit of a sum of reversible Carnot cycles.

In order to utilize the entropy functions to analyze the spontaneity of chemical reactions it is necessary to recognize a "fact" which is not formally stated to be a "law." For lack of a better name, we call it the $2\frac{1}{2}$th law. It is related to the *Clausius inequality*

$$\int_{\text{cycle}} \frac{dq}{\text{T}} < 0 \tag{II-44}$$

for a cycle along which T is defined. The fact is just this: an engine operating irreversibly (with frictional losses, etc.) always operates less efficiently than a reversible engine, where both engines operate between the same two temperatures.

We have shown that for cycles operating between two fixed temperatures, the second law implies that

$$e \text{ any cycle} \leq e \text{ rev. cycle} \tag{II-45}$$

We here assert that the *equality* is possible only if the cycle is reversible, i.e.,

$$e \text{ irrev. cycle} < e \text{ rev. Carnot cycle} \tag{II-46}$$

Thus

$$\left(\frac{q_2 + q_1}{q_2}\right)^{\text{irrev.}} < \frac{\text{T}_2 - \text{T}_1}{\text{T}_2} \tag{II-47}$$

where the right hand side of this inequality is the efficiency of a Carnot cycle operating between the two "reservoir" temperatures T_1 and T_2 of the irreversible cycle. By applying the same argument used above to obtain the entropy function together with this inequality we obtain

$$\int_{\text{cycle}} \frac{dq_{\text{irrev.}}}{T^{\text{res}}} < 0 \qquad \text{(II-48)}$$

where T^{res} is the temperature of the heat source or sink. Equation II-48 is the **Clausius inequality.**

Now consider an arbitrary spontaneous process which occurs in an adiabatically isolated system such that it takes the system from state A to state B. If the system is then removed from isolation and reversibly returned to its initial state A we have, by equation (II-48)

$$\int_{\text{cycle}} \frac{dq}{T} < 0 \qquad \text{(II-49)}$$

Separating the integral into two parts,

$$\int_{\text{cycle}} \frac{dq}{T} = \int_A^B \frac{dq^{\text{irrev.}}}{T} + \int_B^A \frac{dq^{\text{rev.}}}{T} < 0 \qquad \text{(II-50)}$$

and, since the spontaneous process occurred adiabatically,

$$\int_A^B \frac{dq^{\text{irrev.}}}{T} = 0 \qquad \text{(II-51)}$$

Thus

$$\int_B^A \frac{dq^{\text{rev.}}}{T} < 0 \qquad \text{(II-52)}$$

But reversing the order of integration simply changes the sign of the integral, hence

$$\int_A^B \frac{dq^{\text{rev.}}}{T} > 0 \qquad \text{(II-53)}$$

or

$$\Delta S > 0 \qquad \text{(II-54)}$$

for any change of state which can occur spontaneously in isolation. This is an extremely useful fact about the entropy of an isolated system, namely, that it increases during all spontaneous processes. Furthermore, since the above argument holds in reverse, if the entropy of an isolated system increases then a spontaneous process has occurred.

This fact is useful because it provides a criterion for determining whether a process (chemical or physical) tends to go, i.e., is spontaneous. We consider the *total* entropy change, i.e., the entropy change of *all parts* of a system during the process. Since the system and surroundings are isolated from other systems, by definition, this total entropy change enables us to determine whether or not the process is spontaneous.

It is very important that the total entropy change of the system, and not just the entropy change of one part of the system, be considered. For example, consider the entropy changes incurred when a mole of copper at 205°K is placed in contact with an infinitely (for practical purposes) large heat sink at 200°K. The block, initially at 205°K is cooled to 200°K. The entropy change of the block being a property change, depends only on the initial and the final states of the copper, and thus may be calculated by considering a reversible change between the same initial and final states, namely

$$\Delta S_{\text{block}} = \int_{205}^{200} \frac{dq}{\mathrm{T}} = \int_{205}^{200} \frac{C_p\, d\mathrm{T}}{\mathrm{T}} \qquad \textbf{(II-55)}$$

Using standard integration formulae, we have

$$\int_{x_1}^{x_2} \frac{dx}{x} = \log_e \frac{x_2}{x_1} = \ln \frac{x_2}{x_1} = 2.303 \log_{10} \frac{x_2}{x_1} \qquad \textbf{(II-56)}$$

and thus obtain

$$\Delta S_{\text{block}} = C_p \ln \frac{200}{205} = 2.303 C_p \log \frac{200}{205} \qquad \textbf{(II-57)}$$

$$= -0.024 C_p \text{ cal/deg} \qquad \textbf{(II-58)}$$

a *negative number!* That is, we have obtained a negative entropy change for a spontaneous process! The catch, of course, is that the entire *system* must be considered, and we have not considered the reservoir. The entropy change for the reservoir is simply

$$\Delta S_{\text{reservoir}} = \frac{q}{\mathrm{T}} = C_p{}^{\text{block}} \frac{(\mathrm{T}_2 - \mathrm{T}_1)}{\mathrm{T}_2} = +.025\, C_p$$

a positive number larger in magnitude than ΔS_{block}.

We close this section with an important example of a calculation of a change in entropy. It is most important to realize that, since entropy is a property, the path connecting the initial and final states is irrelevant to the entropy change, which depends only on the states themselves. Thus any process, reversible or irreversible, which takes one mole of an ideal gas from the state defined by P_1, $\overline{V}_1$ and T_1 to the state defined by P_2, $\overline{V}_2$ and T_2 is accompanied by the same entropy change.

We can calculate the entropy change by considering a reversible path connecting the same two states. We note that since the pressure is given by the volume and temperature and the equation of state we need only consider the change from $\overline{V}_1$, T_1 to $\overline{V}_2$, T_2 and the pressure will change accordingly. For convenience, we then split the change into two parts: a constant volume temperature change and an isothermal volume change.

During a constant volume temperature change we have

$$dq^{\text{rev.}} = C_v\, d\text{T} \tag{II-59}$$

and thus

$$dS = \frac{C_v\, d\text{T}}{\text{T}} = C_v d \ln \text{T} \tag{II-60}$$

or

$$\Delta S = \int_{\text{T}_1}^{\text{T}_2} C_v d \ln \text{T} \tag{II-61}$$

For a monatomic ideal gas $C_v = \frac{3}{2}\text{R}$ and

$$\Delta S = \tfrac{3}{2}\text{R} \ln \text{T}_2/\text{T}_1 \tag{II-62}$$

We prove in the appendix that the internal energy of an *ideal gas* is independent of the volume of the gas at constant temperature. Thus for the second part of this process, the isothermal expansion of the gas (state changes from $\overline{V}_1$, T_2 to $\overline{V}_2$, T_2),

$$\Delta U = q + w = 0 \tag{II-63}$$

or

$$q = -w \tag{II-64}$$

We have, for a reversible process,

$$w^{\text{rev.}} = -\int P_{\text{sys}}\, dV = -\frac{\text{RT}_2\, dV}{V} \tag{II-65}$$

or

$$w^{\text{rev.}} = -\text{RT}_2 \ln \frac{V_2}{V_1} \tag{II-66}$$

Substituting this result into equation (II-66) yields

$$q^{\text{rev.}} = \text{RT}_2 \ln \frac{V_2}{V_1} \tag{II-67}$$

and thus

$$\Delta S = \frac{q^{\text{rev.}}}{\text{T}} = \text{R} \ln V_2/V_1 \qquad \text{(II-68)}$$

The total entropy change for the change of state is the sum of the values given in equations (II-68) and (II-61) and thus, for the change of state from $\overline{V}_1$, T_1 to $\overline{V}_2$, T_2

$$\Delta S = \int_{\text{T}_1}^{\text{T}_2} C_v d \ln \text{T} + \text{R} \ln V_2/V_1 \qquad \text{(II-69)}$$

for any ideal gas, and for a monatomic ideal gas

$$\Delta S = \tfrac{3}{2}\text{R} \ln \text{T}_2/\text{T}_1 + \text{R} \ln V_2/V_1 \qquad \text{(II-70)}$$

This calculation has provided an important result for ideal gases. What is more important is that it has illustrated the general method by which one calculates entropy changes, namely by finding a convenient reversible path connecting the initial and final states, and calculating the entropy change along that path.

GIBBS FREE ENERGY

The fact that many chemical reactions of interest occur at constant pressure, and with the same initial and final temperatures, provides a convenient simplification for determining whether the chemical process in question is reversible, or whether it occurs spontaneously (i.e., is irreversible) in a given direction. We have shown that at constant temperature and pressure

$$\Delta H = q \qquad \text{(II-71)}$$

and it follows from the definition of entropy that

$$\text{T}\,\Delta S = q^{\text{rev.}} \qquad \text{(II-72)}$$

if the process in question is carried out isothermally. Thus, for a given process at constant T and P, we can compare the heat actually absorbed, namely ΔH, with the heat that would be adsorbed if the process were carried out reversibly, namely T ΔS. We see immediately from the previous development that

$$\Delta H = \text{T}\,\Delta S \qquad \text{(II-73)}$$

for a process at constant T and P if, and only if, the process is a reversible one, i.e., if at each step of the process it could be reversed by an infinitesimal change in the pressure, temperature or some other property.

Thus we have a criterion for determining whether or not a process carried out at constant T and P is reversible. But if it is not reversible in which direction does it tend to go? To answer this question it is necessary to consider the difference between ΔH and $\mathrm{T}\,\Delta S$. The function $H - \mathrm{T}S$ is called the **Gibbs free energy,** G, i.e.,

$$G = H - \mathrm{T}S \tag{II-74}$$

It is clear that G is a property since H, T and S are all properties, and has the units of energy. At constant temperature

$$\Delta G = \Delta H - \mathrm{T}\,\Delta S \tag{II-75}$$

For any change in state of a system which can be carried out reversibly at constant temperature and pressure $q_p = \Delta H = \mathrm{T}\,\Delta S$, where the temperature is that of the initial and final states (necessarily these are equal) and thus $\Delta G = 0$. Conversely, if for two states $\Delta G = 0$ then the constant temperature-constant pressure process between the two states is reversible.

Furthermore, for a spontaneous process at constant temperature and pressure there is a relationship between the sign of ΔG and the direction in which the process tends to go. To see this, consider any process which occurs at constant T and P which takes a system from state 1 to state 2. For this process

$$\Delta G = \Delta H - \mathrm{T}\,\Delta S \tag{II-76}$$

and

$$\begin{aligned} \Delta H &= \Delta U + \Delta(PV) \\ &= q + w + P\,\Delta V \end{aligned} \tag{II-77}$$

If the process were carried out reversibly

$$q = \mathrm{T}\,\Delta S \tag{II-78}$$

and

$$\Delta H = \mathrm{T}\,\Delta S + w + P\,\Delta V \tag{II-79}$$

Substituting this into the expression given above for ΔG yields

$$\Delta G = w + P\,\Delta V \tag{II-80}$$

We can separate the work into two kinds, namely $P - V$ work, given by (recall $P_{\mathrm{ex}} = P_{\mathrm{sys}} = $ constant)

$$w_{P-V} = -\int P_{\mathrm{ex}}\,dV = -P\,\Delta V \tag{II-81}$$

and other work w' given by

$$\begin{aligned} w &= w' + w_{P-V} \\ &= w' - P\Delta V \end{aligned} \tag{II-82}$$

Substituting equation (II-82) into equation (II-80) yields

$$\Delta G = w' \tag{II-83}$$

Thus ΔG is the work other than $P - V$ work done on the system when the process is carried out *reversibly*. This is the minimum amount of work which we must put into the system to make the constant temperature-constant pressure process go, and is the negative of the *maximum* work we can obtain from the system at constant temperature and pressure. Clearly, if we must put work into the system ($w' > 0$), the process is not spontaneous, and if we can obtain work from the system ($w' < 0$) the process is spontaneous. Thus

$$\Delta G < 0 \tag{II-84}$$

implies a spontaneous process,

$$\Delta G = 0 \tag{II-85}$$

implies a reversible process, and

$$\Delta G > 0 \tag{II-86}$$

implies that the reverse process is spontaneous, where the process is that which takes the system from state 1 to state 2 with $w' = 0$.

SUMMARY

In this chapter we have stated the second law of thermodynamics and demonstrated that this law implies the existence of the property called entropy. We have demonstrated that entropy provides a criterion for the spontaneity of processes in isolation. We have illustrated the technique of calculation of entropy changes using the ideal gas equation of state. Finally, we have defined the Gibbs free energy and demonstrated that for a process at constant temperature and pressure the Gibbs free energy change provides a criterion for spontaneity. The important general equations developed in this chapter are

$$dS = dq^{\text{rev.}}/\text{T}$$

$$\int_{\text{cycle}} \frac{dq^{\text{irrev.}}}{\text{T}} < 0$$

$$\Delta G = \Delta H - \text{T}\,\Delta S \qquad (\text{T}, P \text{ constant})$$

Problems

II-1 Calculate the entropy change for the bullet hitting the block in problem I-1 if the bullet and block are initially at 298°K.

II-2 A high temperature, gas-cooled reactor might provide 1000 megawatts of thermal power at a heat source (He gas) temperature of 800°C. What is the maximum work power that can be obtained from such a reactor? What might be done to increase the maximum efficiency of such energy conversion in the future?

II-3 Demonstrate that if it were possible to take a system from state A to state B along two different paths, one of which was a reversible adiabatic path and the other of which was an isothermal path, then it would be possible to convert quantitatively heat into work, and thus to violate the second law.

II-4 Using the fact that the minimum amount of non $P - V$ work required to accomplish the process of lifting one mole of an ideal gas a height h is given by Mgh, obtain the barometric formula

$$P = P^{\circ}e^{-\frac{Mgh}{\text{RT}}}$$

which relates the pressure of a gas at a height h, P, to the pressure at sea level, P°. The enthalpy, like the internal energy, of an ideal gas is independent of volume at constant temperature. Assume that the temperature of the gas does not vary with h.

CHAPTER III

Chemical Equilibrium and The Second Law

THE CLAPEYRON EQUATION

We now find it necessary to make further use of differential and integral calculus. We do so without the type of introduction and justification of formulae given in Chapter I.

Some very useful relationships among the properties V, H, G, S, T, P, etc. result from strictly mathematical manipulation of the *Gibbsian Equations*. The Gibbsian equation for U in a closed system of constant composition is

$$dU = \mathrm{T}\,dS - P\,dV \tag{III-1}$$

and is obtained by noting that for a system in a given state

$$dq^{\text{rev.}} = \mathrm{T}\,dS \tag{III-2}$$

and

$$dw^{\text{rev.}} = -P_{\text{sys.}}\,dV \tag{III-3}$$

and, by the first law,

$$dU = dq^{\text{rev.}} + dw^{\text{rev.}} \tag{III-4}$$

Since U, T, P and V are properties the equality (III-1) holds for a system in a given equilibrium state. We have

$$G = H - \mathrm{T}S \tag{III-5}$$

and

$$H = U + PV \tag{III-6}$$

by the definitions of G and H. Hence

$$G = U + PV - \mathrm{T}S \tag{III-7}$$

and

$$dG = dU + P\,dV + V\,dP - \mathrm{T}\,dS - S\,d\mathrm{T} \qquad \textbf{(III-8)}$$

where the chain rule of differentiation has been applied. Substitution for dU yields

$$dG = -S\,d\mathrm{T} + V\,dP \qquad \textbf{(III-9)}$$

the Gibbsian equation for G applicable to closed systems with constant composition.

An immediate application of equation (III-9) is to the analytical description of lines in a one component "phase diagram," such as illustrated in Fig. 3.1 for water. The interpretation of the figure is

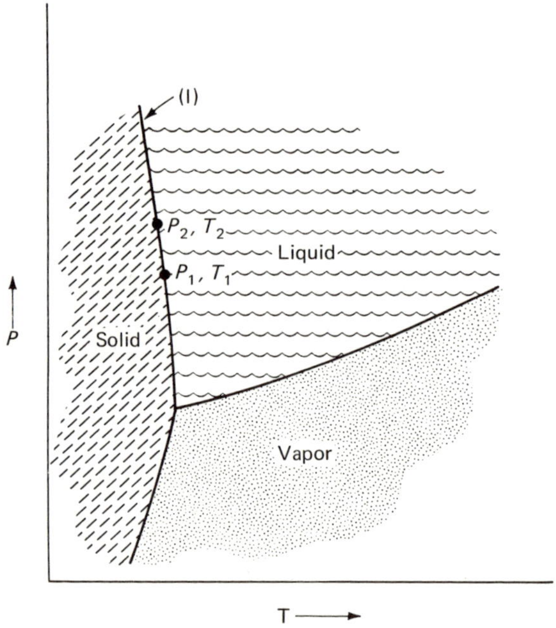

Figure 3.1 A one component phase diagram of water.

simply that at any point in the diagram which corresponds to a particular temperature and pressure the state of the system is as specified. Thus the figure pictorially tells one what the physical state of water is at all temperatures and pressures included on the diagram. The line between the regions labeled liquid and solid, for example, describes the temperatures and pressures at which liquid and solid co-exist in equilibrium. It is just these lines between two phases in equilibrium that are described by the condition

$$\Delta G|_{T,P} = 0 \tag{III-10}$$

For example, consider the line labeled (I) separating the regions labeled liquid and solid on Fig. 3.1. At T_1 and P_1 $\Delta G = 0$ for the process $H_2O(l) = H_2O(s)$. At T_2 and P_2 $\Delta G = 0$ for the same process. Therefore along the line T_1, P_1 to T_2, P_2

$$\bar{G}^{\text{liq}}(T_1, P_1) - \bar{G}^{\text{liq}}(T_2, P_2) = \bar{G}^{\text{solid}}(T_1, P_1) - \bar{G}^{\text{solid}}(T_2, P_2) \tag{III-11}$$

or, for infinitesimal changes of state

$$d(\bar{G}^{\text{liq}} - \bar{G}^{\text{solid}}) = d(\Delta G) = 0 \tag{III-12}$$

But, by equation (III-9)

$$d(\Delta\bar{G}) = -(\Delta\bar{V})\,dP + (\Delta\bar{S})\,dT$$

and, thus using equation (III-12), we have for the slope of the equilibrium line

$$dP/dT = \Delta\bar{S}/\Delta\bar{V} \tag{III-13}$$

the Clapeyron Equation. Since $\Delta\bar{S}$ for a reversible process at constant T and P is $\Delta\bar{H}/T$, we have

$$dP/dT = \Delta\bar{H}/T\,\Delta\bar{V} \tag{III-14}$$

An interesting example of the use of this relation is in calculating the reduction in equilibrium liquid-solid temperature due to an ice skater standing on ice. It is clear from the water phase diagram (Fig. 3.1) that increasing the pressure on solid water at constant temperature tends to liquefaction, i.e., causes the system to traverse the path indicated by the dotted line in Fig. 3.2.

The average skater exerts a pressure above atmospheric pressure of $P_2 - P_1 = 18 \times 10^8$ dynes/cm^2. The density of water at 0°C is 1.00 g/cm^3, and of ice at 0° is 0.9 g/cm^3, leading to a molar volume change between water and ice of $\Delta\bar{V} = \bar{V}_{H_2O} - \bar{V}_{ice} = -1.98$ cm^3/mole. The molar heat of fusion of ice is $\Delta\bar{H} = \bar{H}_{H_2O} - \bar{H}_{ice} = 1{,}440$ cal/mole $= 6.03 \times 10^{10}$ erg/mole. Taking dP/dT to be constant, which is valid for a small temperature interval, and taking T to be 273°K, the freezing point of water at $P = 1$ atmos-

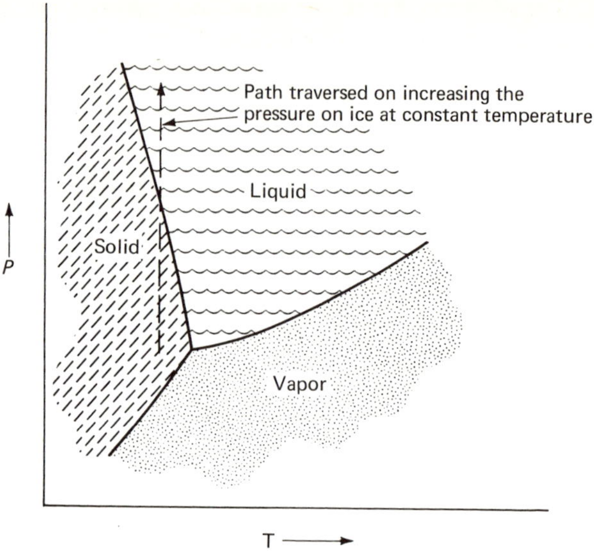

Figure 3.2

phere, we have

$$\Delta T = \frac{T\Delta\overline{V}(P_2 - P_1)}{\Delta\overline{H}} = \frac{273° \times 1.98 \times 18 \times 10^{18}}{6.03 \times 10^{10}} \frac{\text{mole-cm}^3\text{-dyne}}{\text{mole-erg-cm}^2} \quad \text{(III-15)}$$

$$= -16°,* \quad \text{(III-16)}$$

and thus our answer, $-16°$C. This answer will depend, of course on the weight of the skater and the sharpness of his skates.

THE CLAUSIUS-CLAPEYRON EQUATION

In this section we shall obtain the equation of the line separating liquid and vapor on Fig. 3.2. For liquid-vapor equilibrium, $\Delta\overline{V}$ is, to 0.1%, simply equal to $\overline{V}_{gas}$ (molar volumes of gases are typically of the order of tens of liters, and of liquids of the order of tens of cm^3).

As discussed previously, gases at sufficiently high temperatures and low pressures approximately obey the ideal gas equation of state,

* The same result is obtained with no approximation by integrating equation (III-14) to yield $\ln(T_2/T_1) = (P_2 - P_1)\,\Delta\overline{V}/\Delta\overline{H}$.

and thus the molar volume of a gas is approximately given by

$$\overline{V}_{gas} = RT/P \tag{III-17}$$

where

$$\begin{aligned} R &= 0.082 \text{ liter-atm/mole deg} \\ &= 82 \text{ cm}^3\text{-atm/mole deg} \\ &= 1.986 \text{ cal/mole deg} \\ &= 8.315 \text{ joules/mole deg} \end{aligned}$$

the different values resulting from conversion of one set of units to the other.

Thus,

$$dP/dT = \Delta\overline{H}/\Delta\overline{V} = \Delta\overline{H}/T\overline{V}_{gas} = (\Delta\overline{H}/T)(P/RT) \tag{III-18}$$

By mathematics

$$dP/P = d\,(\ln P) \tag{III-19}$$

and

$$dT/T^2 = -d(1/T) \tag{III-20}$$

Combining equations (III-18), (III-19), and (III-20) yields

$$d(\ln P)/d(1/T) = -\Delta\overline{H}_{vap}/R \tag{III-21}$$

Equation III-21 is the second law basis for determining heats of vaporization; a plot of ln P *vs.* $1/T$ will have a slope of $-\Delta\overline{H}_{vap}/R$, as illustrated in Fig. 3.3. While equation (III-21) is, at first glance, not

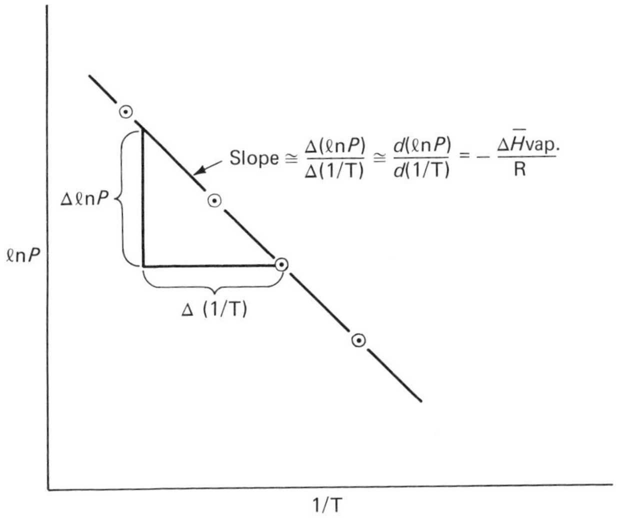

Figure 3.3 Second law heat of vaporization from temperature dependence of vapor pressure.

a relation to send the uninitiated into ecstacy, it is a boon indeed to those who have attempted to measure a "First Law" heat of vaporization by supplying a measured amount of heat to a liquid at its boiling point, and quantitatively collecting the vapor boiled off! Pressure and temperature are much more easily measured than are heat and weight.

Integrating equation III-21 yields

$$\ln (P_2/P_1) = \Delta\overline{H}(T_2 - T_1)/RT_2T_1 \qquad \textbf{(III-22)}$$

the explicit relation between temperature, vapor pressure, and heat of vaporization known as the Clausius-Clapeyron Equation.

Boiling Eggs

It is clear from equation (III-22) why a two-minute egg in New York City will be a four-minute egg in Denver. Since atomospheric pressure is one atmosphere in New York City (roughly at sea level), and 0.816 atm in Denver, "The Mile High City," water in New York City boils at 373°K, whereas water in Denver boils at a lower temperature, T_2. T_2 is found by substituting appropriate values in equation (III-22).

$$\ln (0.816/1) = \frac{9{,}720 \text{ cal/mole } (T_2 - 373)}{2 \text{ cal/mole deg } (373\ T_2)} \qquad \textbf{(III-23)}$$

This yields a value of T_2 = 364°K, or 91°C, which is 9 degrees lower than in New York City!

CHANGES IN THE GIBBS FREE ENERGY OF AN IDEAL GAS

Table III-1 gives dimensionless values of $P\overline{V}/RT$, where P is the pressure of oxygen at a molar volume $\overline{V}$ and temperature T, and R is the gas constant with appropriate units. This table illustrates that at sufficiently low pressure at a given temperature, the equation $P\overline{V} = RT$ closely approximates the behavior of oxygen. This is true for all gases, although what constitutes a "sufficiently low" pressure varies from gas to gas. This equation is the equation of state of an ideal gas, as discussed in Chapter I. From this relation it is possible to calculate changes in thermodynamic properties for ideal gases (as illustrated below) and thus, since all gases become essentially ideal at low pressures, obtain changes in some thermodynamic properties of real gases at low pressures.

Table III.1

Values of PV/RT for Oxygen Calculated from the Beattie-Bridgeman Equation

	P, atmospheres											
T, °K	0.1	0.5	1	2	5	10	20	50	100	200	500	1000
100	0.99776	0.98870	0.97715	0.95323								
200	0.99969	0.99846	0.99691	0.99382	0.98451	0.96889	0.93731	0.84100	0.70091	0.69090	1.07806	1.67361
300	0.99994	0.99967	0.99935	0.99870	0.99677	0.99362	0.98761	0.97189	0.95454	0.95540	1.12621	1.50296
400	1.00000	1.00000	1.00000	1.00001	1.00003	1.00010	1.00038	1.00225	1.00871	1.03290	1.16180	1.42812
500	1.00002	1.00012	1.00023	1.00047	1.00117	1.00236	1.00478	1.01249	1.02663	1.05902	1.17588	1.38877
600	1.00003	1.00016	1.00032	1.00064	1.00160	1.00321	1.00644	1.01629	1.03313	1.06809	1.17808	1.36183
700	1.00004	1.00018	1.00035	1.00070	1.00175	1.00351	1.00702	1.01757	1.03518	1.07042	1.17474	1.34006
800	1.00004	1.00018	1.00036	1.00071	1.00178	1.00357	1.00712	1.01775	1.03531	1.06984	1.16895	1.32115
900	1.00004	1.00018	1.00035	1.00070	1.00176	1.00351	1.00701	1.01743	1.03456	1.06795	1.16222	1.30425
1000	1.00003	1.00017	1.00034	1.00068	1.00171	1.00341	1.00680	1.01689	1.03342	1.06551	1.15527	1.28897

Reprinted from Los Alamos Scientific Laboratory report number LA-2271, by C. E. Holley, Jr., *et al.*, with permission of the authors.

Let us then examine changes in various properties when an ideal gas is expanded or compressed at *constant temperature.* Thermodynamic arguments given in Appendix II show that the enthalpies and internal energies of ideal gases do not change unless the temperature of the gas is changed. Thus

$$(\Delta \overline{U})_T = (\Delta \overline{H})_T = 0 \qquad \text{(III-24)}$$

and since

$$\Delta \overline{G} = \Delta \overline{H} - T\Delta \overline{S} \qquad \text{(III-25)}$$

we have

$$\Delta \overline{G} = -T\Delta \overline{S} \qquad \text{(III-26)}$$

But at constant temperature the change of entropy of an ideal gas is equal to R ln $\overline{V}_2/\overline{V}_1$, as shown by equation (II-70). For an ideal gas at constant temperature we have

$$P_1\overline{V}_1 = P_2\overline{V}_2 \qquad \text{(Boyle's Law)} \qquad \text{(III-27)}$$

or

$$P_1/P_2 = \overline{V}_2/\overline{V}_1 \qquad \text{(III-28)}$$

and thus

$$\Delta \overline{S} = -R \ln P_2/P_1 \qquad \text{(III-29)}$$

and

$$\Delta \overline{G} = RT \ln P_2/P_1 \qquad \text{(III-30)}$$

This important relationship shows how the Gibbs free energy of an ideal gas varies with pressure at constant temperature. It can be used to develop relationships between the pressures of reactants and products of a chemical reaction involving gaseous substances (strictly ideal gaseous substances) at equilibrium. We develop a relationship of this type in the next section of this chapter.

GAS PHASE EQUILIBRIUM

In this section we shall consider the Gibbs free energy change accompanying a chemical reaction in which all reactants and products are ideal gases. In later sections we shall consider more general reactions using the ideas developed here as a stepping stone.

Consider a mixture of $COCl_2$(g) (phosgene), CO(g), and Cl_2(g) in a closed container. These substances are related by the chemical reaction

$$COCl_2(g) = CO(g) + Cl_2(g) \qquad \text{(III-31)}$$

One characteristic of an ideal gas, as is shown in Appendix III, is that its Gibbs free energy in a gaseous mixture is equal to the Gibbs free energy of pure gas at the same partial pressure and temperature. Thus if we consider separating the $COCl_2(g)$ from the mixture and compressing (or expanding) it to 1 atm we obtain the Gibbs free energy change for the process by utilizing equation (III-30), namely

$$\overline{G}^{\circ}_{COCl_2} - \overline{G}_{COCl_2} = RT \ln 1/P_{COCl_2} \tag{III-32}$$

We similarly obtain

$$\overline{G}^{\circ}_{CO} - \overline{G}_{CO} = RT \ln 1/P_{CO} \tag{III-33}$$

for CO(g), and

$$\overline{G}^{\circ}_{Cl_2} - \overline{G}_{Cl_2} = RT \ln 1/P_{Cl_2} \tag{III-34}$$

for $Cl_2(g)$, where, for example $\overline{G}^{\circ}_{Cl_2}$ is the molar Gibbs free energy of pure Cl_2 at 1 atm pressure and $\overline{G}_{Cl_2}$ is the molar Gibbs free energy of the gas in the gaseous mixture at the pressure P_{Cl_2}. Thus we can obtain the Gibbs free energy change per mole of $COCl_2(g)$ at P_{COCl_2} that dissociates to form CO(g) and $Cl_2(g)$ at P_{CO} and P_{Cl_2}, respectively, by combining equations (III-32), (III-33), and (III-34) to obtain

$$\overline{G}_{CO} + \overline{G}_{Cl_2} - \overline{G}_{COCl_2} = \overline{G}^{\circ}_{CO} + \overline{G}^{\circ}_{Cl_2} - \overline{G}^{\circ}_{COCl_2} + RT \ln \frac{P_{CO}P_{Cl_2}}{P_{COCl_2}} \tag{III-35}$$

We call the left hand side of equation (III-35) ΔG, the Gibbs free energy change for the reaction, and similarly define ΔG°. Thus we can rewrite equation (III-35)

$$\Delta G = \Delta G^{\circ} + RT \ln \frac{P_{CO}P_{Cl_2}}{P_{COCl_2}} \tag{III-36}$$

Suppose now that the $COCl_2(g)$, CO(g) and $Cl_2(g)$ are in equilibrium with respect to reaction (III-31). This means that if a piston on the system were moved infinitesimally the reaction would proceed to the left or to the right to produce or consume infinitesimal quantities of reactants or products. Consider disturbing the equilibrium in such a manner that Δn moles of $COCl_2(g)$ are consumed. Let the corresponding Gibbs free energy change be $\Delta G'$. The ratio $\Delta G'/\Delta n$ is the Gibbs free energy change for the reaction, as defined above, and the limit of this ratio as Δn approaches zero is exactly the Gibbs free energy change per mole of $COCl_2$ consumed at equilibrium. But the process at equilibrium is reversible, and thus dG/dn, or the Gibbs free energy change for the process at equilibrium, is zero at constant T and P. Thus if we set ΔG in equation (III-36) equal to zero we shall obtain an equation appropriate to equilibrium. What we obtain is

$$\Delta G^\circ = -\mathrm{RT} \ln \frac{P_{\mathrm{CO}}P_{\mathrm{Cl_2}}}{P_{\mathrm{COCl_2}}} \qquad \textbf{(III-37)}$$

The quantity on the left of this equation is the Gibbs free energy change which accompanies the process when one mole of pure $COCl_2(g)$ at 1 atm pressure and the temperature in question is converted into one mole each of pure CO(g) and $Cl_2(g)$, both pure and each at 1 atm partial pressure and the temperature in question. We define the state of an ideal gas at one atmosphere pressure and the temperature T to be the *standard state* of the ideal gas at T, and thus call ΔG° the **standard Gibbs free energy change for the reaction.** The value of ΔG° is a constant for a given reaction at a given temperature, since the Gibbs free energy of each species in the reaction is a function only of T, P, and n, and all three are specified for the gas in the standard state with $n = 1$.

The right hand side of equation (III-37) contains the gas constant, R, the temperature, T, and the natural logarithm of the ratio $P_{\mathrm{CO}}P_{\mathrm{Cl_2}}/P_{\mathrm{COCl_2}}$. We can solve for this ratio and obtain

$$\frac{P_{\mathrm{CO}}P_{\mathrm{Cl_2}}}{P_{\mathrm{COCl_2}}} = e^{-\frac{\Delta G^\circ}{\mathrm{RT}}} \qquad \textbf{(III-38)}$$

which is constant at a given temperature, i.e.,

$$\frac{P_{\mathrm{CO}}P_{\mathrm{Cl_2}}}{P_{\mathrm{COCl_2}}} = \mathrm{K} \qquad \textbf{(III-39)}$$

This constant is called the *equilibrium constant*, for it provides the relationship between the pressures of the reactants and products at a fixed temperature and at equilibrium, i.e., when $\Delta G = 0$.

Such an expression is useful, for it permits chemists to calculate the extent to which a reaction proceeds when reactants combine to form equilibrium amounts of products in a system. We note here that it is a measure of the tendency for a given reaction to proceed, but that it tells us nothing about the rate at which the reaction will proceed toward equilibrium. This is in general true about thermodynamic results, for thermodynamics deals with interrelationships between initial and final states of rest but not with the path or paths connecting the states. In this text we shall therefore ignore the question of whether a reaction is sufficiently rapid to make it important, and deal only with the question of what the final state in a given reaction system will be if the reaction proceeds to equilibrium.

We shall illustrate the application of thermodynamics to a chemical problem by putting numbers into the equations developed above for the dissociation of phosgene.

For the reaction

$$COCl_2(g) = CO(g) + Cl_2(g) \tag{III-40}$$

at 500°K, $\Delta G° = 39.5$ kcal*, and thus

$$K = e^{-\Delta G°/RT} = 10^{-39,500/4.576 \cdot 500} = 10^{-17.3} \tag{III-41}$$

The 4.576 in the denominator of the exponent of 10 in equation (III-41) is the product of the gas constant, in units of calories per mole per degree, and the natural logarithm of 10:1.987 × 2.303 = 4.576 cal/mole deg.

Returning to the chemical problem, let us suppose that we mix CO(g) and $Cl_2(g)$, both at 1 atm, in a closed container at 500°K. Since the pressure of each gas is proportional to the number of moles of that gas, and since the gases are produced or consumed in the ratios 1:1:1 in reaction (III-40), we can write

$$P_{CO} = 1 - X \tag{III-42}$$

and

$$P_{Cl_2} = 1 - X \tag{III-43}$$

where

$$P_{COCl_2} = X \tag{III-44}$$

and thus obtain

$$(1 - X)^2/X = 10^{-17.3} \tag{III-45}$$

But if the ratio on the left of equation (III-45) is very small, which is clearly the case, then the numerator of the ratio must be very small as well. The numerator is very small only if $X \cong 1$, in which case the ratio is essentially equal to the numerator. Thus

$$(1 - X)^2 \cong 10^{-17.3} \tag{III-46}$$

Combining equations (III-42) and (III-43) with equation (III-46) yields

$$P_{CO} = P_{Cl_2} = 10^{-8.65} \text{ atm} \tag{III-47}$$

Thus we see that it would be feasible to synthesize phosgene by the direct combination of carbon monoxide and chlorine, both at one atmosphere pressure and 500°K, for the amount of reactants remaining after equilibrium is established is negligibly small. On the other hand, if we wish to destroy phosgene at one atmosphere pressure,

* We shall later deal with methods of obtaining values for $\Delta G°$.

we will have to heat it to a higher temperature than 500°K before it dissociates significantly. There are many variations to chemical problems of this type; *e.g.*, the initial pressure of the reactants might differ, the stoichiometric coefficients might differ, and there might be some product initially present. The solutions to the chemical problems can generally be obtained algebraically once the appropriate mole (or mass, or pressure) relationships and appropriate equilibrium constants are obtained such that one has as many equations as unknowns. We shall provide further examples of problems of this sort in Chapter V.

If the reaction involves more than one mole of reactant or product, e.g.,

$$N_2(g) + 3\,H_2(g) = 2\,NH_3(g) \qquad \textbf{(III-48)}$$

we have

$$\Delta G = 2\overline{G}_{NH_3} - 3\overline{G}_{H_2} - \overline{G}_{N_2} = 0 \qquad \textbf{(III-49)}$$

and

$$\Delta G^{\circ} = -RT \ln \frac{P^2_{NH_3}}{P_{N_2} P_{H_2}{}^3} \qquad \textbf{(III-50)}$$

and thus for the equilibrium constant,

$$K = \frac{P^2_{NH_3}}{P_{N_2} P_{H_2}{}^3} \qquad \textbf{(III-51)}$$

THE CHEMICAL POTENTIAL

In order to deal with equilibria in solutions it is necessary that we introduce a new property called the **chemical potential.** As we shall show in more detail in Chapter VII, if we fix the temperature, the pressure, and the number of moles of each component in a solution, we fix the state of the solution. For example, if in a solution of NH_3 in H_2O we fix the number of moles of NH_3 and the number of moles of H_2O, as well as the temperature of and the pressure on the solution, we fix all other properties of the solution. In particular we fix the Gibbs free energy. In mathematical symbols we write for the Gibbs free energy of the solution

$$G = G(T, P, n_{NH_3}, n_{H_2O}) \qquad \textbf{(III-52)}$$

or we say, the Gibbs free energy can be expressed as a function of T, P, n_{NH_3} and n_{H_2O}. This means that the total differential of G

is given by

$$dG = \left.\frac{\partial G}{\partial \mathrm{T}}\right)_{P,n_{\mathrm{NH_3}},n_{\mathrm{H_2O}}} d\mathrm{T} + \left.\frac{\partial G}{\partial P}\right)_{\mathrm{T},n_{\mathrm{NH_3}},n_{\mathrm{H_2O}}} dP + \left.\frac{\partial G}{\partial n_{\mathrm{NH_3}}}\right)_{\mathrm{T},P,n_{\mathrm{H_2O}}} dn_{\mathrm{NH_3}} + \left.\frac{\partial G}{\partial n_{\mathrm{H_2O}}}\right)_{\mathrm{T},P,n_{\mathrm{NH_3}}} dn_{\mathrm{H_2O}} \qquad \text{(III-53)}$$

By comparison with equation (III-9), we note that equation (III-53) is equivalent to

$$dG = -S\,d\mathrm{T} + V\,dP + \overline{G}_{\mathrm{NH_3}}\,dn_{\mathrm{NH_3}} + \overline{G}_{\mathrm{H_2O}}\,dn_{\mathrm{H_2O}} \qquad \text{(III-54)}$$

where we have introduced the symbol $\overline{G}_{\mathrm{NH_3}}$ for the partial derivative of G with respect to the number of moles of ammonia at constant temperature, pressure and number of moles of water, and a similar substitution has been made for the partial of G with respect to the number of moles of water. These partial derivatives are properties of the solution and are the *chemical potentials* of the ammonia and the water, respectively, in the solution.

The significance of the chemical potential is, at first, a little difficult to see. Physically it is the ratio of the change in the Gibbs free energy of the solution to the moles of solute added (NH_3 or H_2O in the case above) in the limit as the number of moles added goes to zero. Note that the Gibbs free energy change goes to zero as well as do the number of moles added, and thus the ratio approaches a noninfinite limiting value. For a pure substance, such as NaCl or H_2O, the ratio of the change in Gibbs free energy to moles added is a constant, and is equal to the molar Gibbs free energy. Thus the chemical potential of a pure substance is simply equal to the molar Gibbs free energy.

In Appendix III it is demonstrated that the chemical potential of an ideal gas depends only upon its partial pressure, and does not depend upon whether or not it is pure or is in a mixture of ideal gases. Thus the Gibbs free energy of an ideal gas and its chemical potential have the same form, and can be said to be the same. Thus we can use the same symbol for the chemical potential as we used for the Gibbs free energies of the gases in a gaseous mixture (in our discussion of gas phase equilibria in the last section). For an ideal gas

$$\overline{G}_{\mathrm{i}} = \overline{G}_{\mathrm{i}}^{\circ} + \mathrm{RT}\ln P_{\mathrm{i}} \qquad \text{(III-55)}$$

i.e., the chemical potential is equal to the chemical potential of the pure gas at one atmosphere and the temperature in question plus RT times the logarithm of the partial pressure of the gas. The validity of this equation for an ideal gas is established in Appendix III.

In the case of solutions, the chemical potential is a most useful property. To illustrate this, let us consider again the solution of NH_3 in H_2O. Let us suppose that we do an experiment in which the solution is separated from the vapor above it by a membrane which is permeable only to NH_3 but not to water. The fact that no such membrane is known to exist need not concern us; the conclusions that we reach on the basis of this thought experiment will be valid. What we do in this thought experiment is to change the volume of the vapor phase, say by the slight movement of a piston. If the volume of the vapor phase is decreased then some NH_3 passes from the vapor phase into the solution phase and the concentration of the solution increases. If the volume is increased it has the opposite effect. The Gibbs free energy changes accompanying these volume changes are given by equation (III-54) for both the solution and the gas phase, provided the system remains in equilibrium. We can therefore apply equation (III-54) to obtain the condition for equilibrium with respect to transfer of ammonia between the solution and the gas phase.

The Gibbs free energy change of the total system is given by equation (III-54) with $dT = dn_{\mathrm{NH}_3}^{\mathrm{total}} = dn_{\mathrm{H}_2\mathrm{O}}^{\mathrm{total}} = 0$ since the process is carried out at constant temperature in a closed system. Thus

$$dG^{\mathrm{total}} = V^{\mathrm{total}} dP \tag{III-56}$$

But for the gas phase, using equation (III-54) and the fact that the number of moles of H_2O in the gas phase and the temperature of the gas phase are constant,

$$dG^{\mathrm{gas}} = V^{\mathrm{gas}}\, dP + \overline{G}_{\mathrm{NH}_3}(\mathrm{gas})\, dn_{\mathrm{NH}_3}^{\mathrm{gas}} \tag{III-57}$$

and similarly for the solution phase

$$dG^{\mathrm{soln.}} = V^{\mathrm{soln.}}\, dP + \overline{G}_{\mathrm{NH}_3}(\mathrm{soln.})\, dn_{\mathrm{NH}_3}^{\mathrm{soln.}} \tag{III-58}$$

But since the Gibbs free energy of the total system is the sum of the free energies of the gas and the solution we have, by (III-56), (III-57) and (III-58),

$$\begin{aligned} V^{\mathrm{total}}\, dP = (V^{\mathrm{gas}} + V^{\mathrm{soln.}})\, dP &+ \overline{G}_{\mathrm{NH}_3}(\mathrm{gas})\, dn_{\mathrm{NH}_3}^{\mathrm{gas}} \\ &+ \overline{G}_{\mathrm{NH}_3}(\mathrm{soln.})\, dn_{\mathrm{NH}_3}^{\mathrm{soln.}} \end{aligned} \tag{III-59}$$

where we have made use of the fact that the pressure on both phases is the same as that on the total system. Since the total volume is the sum of the volumes of the gas and solution phases, we can cancel

$V^{\text{total}}\, dP$ from both sides of equation (III-59) and obtain

$$0 = \overline{G}_{NH_3}(\text{gas})\, dn^{\text{gas}}_{NH_3} + \overline{G}_{NH_3}(\text{soln.})\, dn^{\text{soln.}}_{NH_3} \tag{III-60}$$

We once again make use of the fact that the total system is closed in order to obtain

$$-dn^{\text{gas}}_{NH_3} = dn^{\text{soln.}}_{NH_3} \tag{III-61}$$

i.e., the NH_3 gained by one phase is lost by the other. Combining equations (III-60) and (III-61) yields our desired result:

$$\overline{G}_{NH_3}(\text{gas}) = \overline{G}_{NH_3}(\text{soln.}) \tag{III-62}$$

Thus if the solution and the gas phase are in equilibrium with respect to transfer of NH_3, then the chemical potential of the ammonia is the same in both phases. This is a perfectly general result, for our derivation in no way depended upon the nature of the phases. We can therefore state that if two phases are in equilibrium with respect to transfer of a substance, then the chemical potential of the substance is the same in both phases. This fact will be particularly useful when you know how chemical potentials vary with the concentration of a solution, the subject matter of the next several sections.

CHEMICAL POTENTIAL CHANGES OF NONIONIC SOLUTE SPECIES

In order to discuss the chemical potential of solute species we introduce a non-thermodynamic empirical relationship known as Henry's Law, just as we introduced the empirical ideal gas law to enable us to find the form for the chemical potential of an ideal gas. As is usual in the thermodynamic approach, we use the results of experimental observations without inquiring into the significance of the relationships obtained. The nature of relationships such as the ideal gas equation of state and Henry's Law depends upon the interactions of the atoms or molecules in the different environments. The nature of such interactions, and even the existence of atoms and molecules, is not relevant to the classical thermodynamic (macroscopic) approach.

Henry's Law relates the "mole fraction" of a solute species to the partial pressure of the species over the solution, and is thought to be valid for all nonionic solutions provided they are sufficiently dilute. The law is simply

$$P = k\text{N} \tag{III-63}$$

where P is the partial pressure of the species over the solution, N is the mole fraction of the species in the solution and k is a function of temperature only, and is called the **Henry's law constant.** The mole fraction of component i in solution is the ratio of the number of moles of i to the total number of moles in the solution. For example, in a two component solution consisting of one mole of benzene and two moles of alcohol, the mole fraction of benzene is

$$N_{\text{benzene}} = \frac{\text{Moles of benzene}}{\text{Total moles}} = \frac{1}{3} \qquad \text{(III-64)}$$

and the mole fraction of alcohol is

$$N_{\text{alcohol}} = \frac{\text{Moles of alcohol}}{\text{Total moles}} = \frac{2}{3} \qquad \text{(III-65)}$$

and, as usual, $N_{\text{benzene}} + N_{\text{alcohol}} = 1$. Solutions in which all solutes obey Henry's law are called **ideal solutions.**

We shall use Henry's law to obtain the form of the chemical potential of a species in an ideal solution by considering changes in the chemical potential of the vapor above the solution. We assume that the vapor is an ideal gas. Let us consider, for the sake of concreteness, two solutions of NH_3 in H_2O, one with mole fraction N_1 and partial pressure of NH_3 over the solution equal to P_1, the other with the corresponding values N_2 and P_2. The change in chemical potential of the gas accompanying compression from P_1 to P_2 at constant temperature is, as shown by equation (III-30),

$$\Delta\overline{G}_{NH_3} = RT \ln P_2/P_1 \qquad \text{(III-66)}$$

But

$$P_2 = kN_2 \qquad \text{(III-67)}$$

and

$$P_1 = kN_1 \qquad \text{(III-68)}$$

Substituting equations (III-67) and (III-68) into equation (III-66) we obtain

$$\Delta\overline{G}_{NH_3} = RT \ln N_2/N_1 \qquad \text{(III-69)}$$

This equation states that the change in the chemical potential of the vapor is given by RT times the natural logarithm of the ratio of the *mole fractions in the solution.* But, by equation (III-62) the chemical potential of the NH_3 in solution must change by the same amount if the vapor phase and the solution phase are to be in equilibrium in both the initial and final states. Thus equation (III-69) also gives the change in the chemical potential of the species in solution at equilibrium. We can express this mathematically by

$$\overline{G}_{NH_3}(N_2) - \overline{G}_{NH_3}(N_1) = RT \ln N_2/N_1 \qquad \text{(III-70)}$$

i.e., the difference between the chemical potentials of two ideal solutions which differ only in concentration, is RT ln N_2/N_1.

As in the case of ideal gases, it is convenient to define a standard state for solute species. The usual state chosen as a standard state for solute species is the state that we *would* reach if we concentrated the solution until the mole fraction of the solute was unity in such a way that the solution obeyed Henry's Law during the concentration. It is thus a hypothetical standard state for real solute species inasmuch as real solutions deviate from Henry's Law behavior at moderate concentrations, and also since many solutions cannot be concentrated beyond a certain concentration without a phase separation occurring. The fact that the standard state is hypothetical need not concern us, however, since we can calculate the change in chemical potential that *would* accompany the change in state from the standard state to the state of interest, namely, by application of equation (III-70)

$$\overline{G}_i = \overline{G}_i^\circ + RT \ln N_i \qquad \text{(III-71)}$$

There is another way of looking at equation (III-71), and that is to say that the chemical potential of a solute in an ideal solution can be separated into two parts—one part, $\overline{G}_i^\circ$, which does not depend upon concentration and another part, RT ln N_i, which explicitly gives the concentration dependence.

An alternative description is frequently given in which Henry's Law is written

$$P_{NH_3} = k m_{NH_3} \qquad \text{(III-72)}$$

where m is the molality of the solution. Equations (III-67) and (III-72) are equivalent at low concentrations since the molality and the mole fraction become proportional to each other in sufficiently dilute solutions. Equation (III-67) is valid over a longer concentration range than is equation (III-72). Utilizing equation (III-72) in a manner entirely analogous to the treatment of equation (III-67), we obtain

$$\overline{G}_{NH_3}(m_2) - \overline{G}_{NH_3}(m_1) = RT \ln m_2/m_1 \qquad \text{(III-73)}$$

Defining the standard state in this case to be the hypothetical one molal solution, we obtain for the general solute species in an ideal solution

$$\overline{G}_i = \overline{G}_i^\circ + RT \ln m_i \qquad \text{(III-74)}$$

Note that we use the same symbol for the chemical potential of the solute in the standard state for both the molality [equation (III-74)] and the mole fraction [equation (III-71)] description, although these clearly must have different values. This is done for convenience, and it must be understood when using tabulated data to what standard state these data refer.

CHEMICAL POTENTIAL CHANGES OF IONIC SOLUTE SPECIES

Until now we have avoided the problem of ionic species in solution, surely a most important one to chemists! We treat this problem formally in a manner similar to that for NH_3 in solution, namely by considering a species which is completely ionized in solution and co-exists with itself in the vapor in a solution-vapor phase equilibrium. HCl is one such species.

Rather than Henry's Law, it is found for HCl that the relation

$$P_{\mathrm{HCl}}^{\mathrm{gas}} = k m_{\mathrm{HCl}}^{2} = k(m_{\mathrm{H}^+})(m_{\mathrm{Cl}^-}) \qquad \textbf{(III-75)}$$

is obeyed in sufficiently dilute solution. That is, it is found that the partial pressure of gaseous HCl over a dilute solution of HCl is proportional to the square of the concentration of the HCl in solution, and the square of the concentration is equal to the product of the concentrations of the ions. We obtain the expression for the chemical potential of HCl in solution in a manner analogous to that used in the previous section to obtain the expression for the chemical potential of NH_3 in solution;

$$\overline{G}_{\mathrm{HCl}} = \overline{G}_{\mathrm{HCl}}^{\circ} + \mathrm{RT} \ln m_{\mathrm{HCl}}^{2} \qquad \textbf{(III-76)}$$

or

$$\overline{G}_{\mathrm{HCl}} = \overline{G}_{\mathrm{HCl}}^{\circ} + \mathrm{RT} \ln (m_{\mathrm{H}^+})(m_{\mathrm{Cl}^-}) \qquad \textbf{(III-77)}$$

where the chemical potential in the standard state again is the chemical potential of the hypothetical one molal solution which obeys the ideal law, equation (III-75) in this case.

In the more general ideal case where an electrolyte dissolves to form, say, *a* cations and *b* anions, for example, $Al_2(SO_4)_3$, which dissolves to form $a = 2$ cations and $b = 3$ anions, we find that

$$\overline{G}_{\mathrm{electrolyte}} = \overline{G}_{\mathrm{electrolyte}}^{\circ} + \mathrm{RT} \ln (m_{\mathrm{cation}}^{a})(m_{\mathrm{anion}}^{b}) \qquad \textbf{(III-78)}$$

By comparison with experimental results, such as will be discussed in Chapter VI, it is found that electrolytes obey equation (III-78) in sufficiently dilute solutions.

STANDARD STATES AND ACTIVITIES

Since $\overline{G}°$ or $\overline{G}_i°$, the molar Gibbs free energy or chemical potential, respectively, in the standard state, occurs in *all* expressions for free energies of species, it is necessary to understand clearly what is meant by standard state for all species under consideration. Therefore in this section we shall discuss the usual standard states chosen for species of various kinds, although some of what we say will simply be a review of what has preceded.

(1) Ideal Gas. We defined the standard state for an ideal gas to be the pure gas at one atmosphere pressure and the temperature in question. This choice was simply a matter of convenience. It simplified the expression

$$\overline{G}_i = \overline{G}_i° + RT \ln P_i/P° \quad \textbf{(III-79)}$$

where $P°$ is the pressure of the gas in the standard state, to

$$\overline{G}_i = \overline{G}_i° + RT \ln P_i \quad \textbf{(III-80)}$$

This choice allowed us to obtain the equilibrium constants

$$K = e^{-\frac{\Delta G_i°}{RT}} \quad \textbf{(III-81)}$$

for reactions of ideal gases as expressions which did not contain numerically the pressure of the gases in their standard states. That such equilibrium constants are dimensionless stems from the fact that the argument of the natural logarithm in equation (III-80) is in reality the *ratio* of P_i to one atm. For this reason, the pressures calculated when standard tabulated data are used *must* have the units of atmospheres!

(2) Ideal Nonionic Solute Species. For ideal solutions we chose the $m = 1$ (or $N = 1$) ideal solutions as standard states for the same reason as given above, namely the convenience of obtaining an equilibrium constant which contains, at least numerically, only the molalities (or the mole fractions) of the species. (This will be demonstrated in later sections and chapters.) We obtained

$$\overline{G}_i = \overline{G}_i° + RT \ln m_i \quad \textbf{(III-82)}$$

or

$$\overline{G}_i = \overline{G}_i° + RT \ln N_i \quad \textbf{(III-83)}$$

depending upon whether the molality (practical) description or the mole fraction (rational) description was used. It should be quite clear that equations (III-82) and (III-83) do not imply $m_i = N_i$, since the numerical values of $\overline{G}_i°$ in the two equations differ by RT ln ff, where ff is the finagle factor which converts molality to mole fraction. While some authors use different symbols to denote different standard state molar free energies, e.g., G_i° and $G_i^{\circ *}$, we feel that the proliferation of symbols at this level of presentation is annoying and unnecessary as long as the reader understands that each concentration description implies different values for the standard state free energy.

(3) Real Nonionic Solute Species. In the case of real (rather than ideal) solutions the same solute standard states are used as in case 2 above. The prescription for taking a real solution to its standard state is: (a) dilute the solution until it is sufficiently dilute that it obeys the ideal solution equation relating the chemical potential to the concentration and, (b) concentrate the solution along the hypothetical path for which the ideal equation is obeyed until the standard state is reached. Presumably, and it has always turned out to be the case, we can measure the change in chemical potential along the real part of the path (part a) and calculate the change in chemical potential along the hypothetical path (part b) by either equation (III-82) or (III-83). What we find is that for real solutions at moderate and high concentrations the concentration dependence of the chemical potential no longer has the form of either equation (III-82) or (III-83). We are at liberty to express this concentration dependence in any way we wish, e.g., an additive function of concentration or a multiplicative one, etc. What is customarily done is to introduce the *activity coefficients* γ or f, which are functions of concentration only, and obtain

$$\overline{G}_i = \overline{G}_i^\circ + \mathrm{RT} \ln \gamma_i m_i \qquad \text{(III-84)}$$

or

$$\overline{G}_i = \overline{G}_i^\circ + \mathrm{RT} \ln f_i \mathrm{N}_i \qquad \text{(III-85)}$$

The reason that these activity coefficients are introduced in the way that they are is to maintain the form of the chemical potential expression. The reason that γ and f are called activity coefficients is that the argument of the logarithm in the chemical potential expression is generally called the **activity** of the species. We express our belief that solutions become ideal in the limit of extreme dilution as follows:

$$\lim_{m \to 0} \gamma = 1 \qquad \text{(III-86)}$$

and

$$\lim_{\mathrm{N} \to 0} f = 1 \qquad \text{(III-87)}$$

i.e., the activity approaches the molality, or mole fraction, depending upon the description used, as the solution becomes dilute.

(4) Ionic Solutions. In the case of solutions of electrolytes the expression for concentration dependence of the chemical potential becomes

$$\overline{G}_i = \overline{G}_i^\circ + RT \ln (\gamma_{H^+} \cdot m_{H^+} \cdot \gamma_{Cl^-} \cdot m_{Cl^-}) \qquad \text{(III-88)}$$

where we have arbitrarily introduced activity coefficients for both H^+ and Cl^- and have used the practical description, since this description is invariably used in the discussion of ionic species. The equation is seldom used in this form, for single ion activities are not available. We use instead the *mean activity coefficient* which is readily obtained; this is the geometric mean of the ionic activity coefficients. Formally we write

$$\overline{G}_i = \overline{G}_i^\circ + RT \ln \gamma_\pm{}^2 m_\pm{}^2 \qquad \text{(III-89)}$$

where $\gamma_\pm$ for HCl is the geometric mean of the activity coefficients of H^+ and Cl^-:

$$\gamma_\pm = \sqrt{\gamma_{H^+}\gamma_{Cl^-}} \qquad \text{(III-90)}$$

The mean molarity is similarly defined. The mean activity coefficients are measurable since there are means of obtaining the changes in the chemical potential in the solution, such as by measuring the changes in the partial pressure of HCl(g) over the solution. Thus, in principle, the difference between the chemical potential of HCl in solution at some concentration and the chemical potential of solute HCl in its standard state can be obtained by the measurement of the pressure of HCl(g) over the solution. Experimentally there is a more convenient method of obtaining changes in the chemical potentials of electrolytes in solution, namely by electrical measurements. This method will be dealt with in Chapter VI.

In the case of electrolytes which are not 1:1 electrolytes, such as $Al_2(SO_4)_3$, the mean activity coefficient and mean molality are, as above, the geometric means of the quantities for the ions. The expression for the chemical potential has the form

$$\overline{G}_i = \overline{G}_i{}^\circ + RT \ln (\gamma_{Al^{3+}}^2 m_{Al^{3+}}^2 \gamma_{SO_4{}^{2-}}^3 m_{SO_4{}^{2-}}^3) \qquad \text{(III-91)}$$

or

$$\overline{G}_i = \overline{G}_i^\circ + RT \ln \gamma_\pm{}^5 m_\pm{}^5 \qquad \text{(III-92)}$$

(5) Solvents and Pure Condensed Phases. For pure liquids, pure solids, and solvents, we take the standard state to be the liquid or solid at the temperature in question and at a total applied pressure of one atmosphere. Since the chemical potential of a condensed phase varies only a little with pressure in the region normally encountered (say, between zero and ten atmospheres) there is no error introduced by assuming that a condensed phase at a pressure in this range is in its standard state. Thus we write, for pure liquid benzene and pure solid NaCl, for example,

$$\overline{G}_i(\text{Benzene}) = \overline{G}_i^\circ(\text{Benzene}) = \overline{G}^\circ(\text{Benzene}) \qquad \textbf{(III-93)}$$

and

$$\overline{G}_i(\text{NaCl}) = \overline{G}_i^\circ(\text{NaCl}) = \overline{G}^\circ(\text{NaCl}) \qquad \textbf{(III-94)}$$

respectively.

(6) Real Gases. We mention in closing this section that real gases do not obey the ideal gas equation of state at high pressures and/or low temperatures (see Table III-1). To account for this we introduce another activity coefficient, designated the **fugacity coefficient,** (f). This coefficient is defined by

$$\overline{G}_i\ (\text{real gas}) = \overline{G}_i^\circ\ (\text{real gas}) + \text{RT} \ln fP \qquad \textbf{(III-95)}$$

where the standard state is that of the hypothetical ideal gas at one atm pressure and the temperature in question. For any real gas

$$\lim_{P \to 0} f = 1 \qquad \textbf{(III-96)}$$

i.e., real gases approach ideality in the low pressure limit.

CHEMICAL EQUILIBRIUM

We are now in a position to consider the application of the change in chemical potential accompanying a chemical reaction to the most general kind of chemical reaction. Before proceeding to a more complicated example, however, we shall complete what we have started, and consider the two chemical reactions

$$NH_3(\text{soln.}) = NH_3(\text{g}) \qquad \textbf{(III-97)}$$

and

$$HCl(\text{soln.}) = HCl(\text{g}) \qquad \textbf{(III-98)}$$

In the first case we have shown that at equilibrium,

$$\overline{G}_{NH_3}(\text{soln.}) = \overline{G}_{NH_3}(\text{g}) \qquad \textbf{(III-99)}$$

Furthermore, it follows from the discussion of the preceding section that

$$\overline{G}_{NH_3}(\text{soln.}) = \overline{G}^\circ_{NH_3}(\text{soln.}) + \text{RT} \ln \gamma m \qquad \textbf{(III-100)}$$

and

$$\overline{G}_{NH_3}(\text{g}) = \overline{G}^\circ_{NH_3}(\text{g}) + \text{RT} \ln fP \qquad \textbf{(III-101)}$$

Combining equations (III-100) and (III-101) according to equation (III-99) yields

$$\overline{G}^{\circ}_{NH_3}(g) + RT \ln fP = \overline{G}^{\circ}_{NH_3}(soln.) + RT \ln \gamma m \qquad \text{(III-102)}$$

which can be rearranged to yield

$$\overline{G}^{\circ}_{NH_3}(g) - \overline{G}^{\circ}_{NH_3}(soln.) = -RT \ln fP/\gamma m \qquad \text{(III-103)}$$

or

$$fP/\gamma m = e^{-\frac{\Delta G^{\circ}_{NH_3}}{RT}} \qquad \text{(III-104)}$$

This equation says that, at equilibrium, the ratio $fP/\gamma m$ is a constant equal to the negative exponential of the difference between the chemical potential of NH_3 in the gas and solution standard states, divided by RT. This constant (at constant temperature and, strictly, pressure) is the equilibrium constant for reaction (III-97), and is given the symbol K. Thus we can write

$$fP/\gamma m = K \qquad \text{(III-105)}$$

If the ammonia gas behaves ideally at the temperature and pressure in question, then the fugacity coefficient, f, is unity, and if the solution behaves ideally (obeys Henry's law), then the activity coefficient, γ, is unity and

$$P/m = k \qquad \text{(III-106)}$$

We see that thermodynamics simply restates Henry's law and provides a relationship between the Henry's law constant, k, and ΔG_i°, namely

$$k = e^{-\frac{\Delta G_i^{\circ}}{RT}} \qquad \text{(III-107)}$$

In an analogous way we obtain for the HCl solution reaction

$$fP/\gamma_{\pm}^{2} m_{\pm}^{2} = e^{-\frac{\Delta G_i^{\circ}}{RT}} \qquad \text{(III-108)}$$

and, for the ideal gas- ideal solution equilibrium

$$P/m_{\pm}^{2} = k \qquad \text{(III-109)}$$

a restatement of equation (III-75)

For a more complicated example we consider the reaction of hydrochloric acid in aqueous solution with pure liquid mercury to form mercurous chloride (Hg_2Cl_2) and hydrogen gas. We write the balanced net reaction

$$2\,Hg(l) + 2\,HCl(m_{HCl}) = Hg_2Cl_2(m_{Hg_2Cl_2}) + H_2(g) \qquad \text{(III-110)}$$

We have, at constant temperature and pressure [cf. equation (54)]

$$dG = \overline{G}_{H_2}\, dn_{H_2} + \overline{G}_{Hg_2Cl_2}\, dn_{Hg_2Cl_2} + \overline{G}_{HCl}\, dn_{HCl} + \overline{G}_{Hg}\, dn_{Hg} \tag{III-111}$$

But the dn_i values are related by the stoichiometry of the chemical reaction, namely

$$dn_{H_2} = dn_{Hg_2Cl_2} = -\tfrac{1}{2}\, dn_{HCl} = -\tfrac{1}{2}\, dn_{Hg} \tag{III-112}$$

and therefore we can eliminate all other dn_i's in favor of dn_{H_2} in equation (III-111) and obtain

$$dG/dn_{H_2} = \overline{G}_{H_2} + \overline{G}_{Hg_2Cl_2} - 2\,\overline{G}_{HCl} - 2\,\overline{G}_{Hg} \tag{III-113}$$

By the criterion for reversibility of the chemical reaction,

$$dG/dn_{H_2} = 0 \tag{III-114}$$

if the system is at equilibrium with respect to this chemical reaction. Therefore

$$\Delta\overline{G}_i = \overline{G}_{H_2} + \overline{G}_{Hg_2Cl_2} - 2\,\overline{G}_{HCl} - 2\,\overline{G}_{Hg} = 0 \tag{III-115}$$

at equilibrium. In the preceding section we presented the forms for the chemical potentials of each of the species in the reaction, namely

$$\begin{aligned} \overline{G}_{H_2} &= \overline{G}^{\circ}_{H_2} + RT \ln P_{H_2} \\ \overline{G}_{Hg_2Cl_2} &= \overline{G}^{\circ}_{Hg_2Cl_2} + RT \ln \gamma_{Hg_2Cl_2} m_{Hg_2Cl_2} \\ \overline{G}_{HCl} &= \overline{G}^{\circ}_{HCl} + RT \ln \gamma_{\pm}{}^{2} m_{\pm}{}^{2} \\ \overline{G}_{Hg} &= \overline{G}^{\circ}_{Hg} \end{aligned} \tag{III-116}$$

We combine these according to equation (III-115) to obtain

$$\Delta\overline{G}_i = \Delta\overline{G}_i{}^{\circ} + RT \ln \frac{P_{H_2}\gamma_{Hg_2Cl_2} m_{Hg_2Cl_2}}{\gamma_{\pm}{}^{4} m_{\pm}{}^{4}} \tag{III-117}$$

where we have assumed that the hydrogen gas behaves ideally, that Hg_2Cl_2 exists as a molecular species in solution and that HCl is a strong electrolyte. If the system is at equilibrium with respect to reaction (III-110), then $\Delta\overline{G}_i = 0$, and we have

$$\Delta\overline{G}_i{}^{\circ} = -RT \ln \frac{P_{H_2}\gamma_{Hg_2Cl_2} m_{Hg_2Cl_2}}{\gamma_{\pm}{}^{4} m_{\pm}{}^{4}} \tag{III-118}$$

Thus

$$K = \frac{P_{H_2}\gamma_{Hg_2Cl_2} m_{Hg_2Cl_2}}{\gamma_{\pm}{}^{4} m_{\pm}{}^{4}} \tag{III-119}$$

is the equilibrium constant for the reaction. We frequently abbreviate this expression by introducing the symbol a_i for the activities of the species and obtain, in the case under discussion

$$K = \frac{a_{H_2} a_{Hg_2Cl_2}}{a_{HCl}^2} \tag{III-120}$$

For the case of a general chemical reaction

$$bB + cC + \ldots = lL + mM + \ldots \tag{III-121}$$

we obtain, using the same reasoning as above,

$$\Delta G_i = \Delta G_i^\circ + RT \ln \frac{a_B^{\,b} a_C^{\,c} \cdots}{a_L^{\,l} a_M^{\,m} \cdots} \tag{III-122}$$

or, since at equilibrium $\Delta G_i = 0$,

$$K = \frac{a_B^{\,b} a_C^{\,c} \cdots}{a_L^{\,l} a_M^{\,m}} = e^{-\frac{\Delta G_i^\circ}{RT}} \tag{III-123}$$

THE USE OF TABULATED DATA

In the problems at the end of this chapter we have tabulated the standard enthalpies of formation and entropies* of various species involved in chemical reactions. These are the data one obtains from a number of standard sources of thermodynamic information, such as some of the sources listed in Appendix I. In order to use the enthalpies of formation and entropies to obtain equilibrium constants these quantities are combined according to

$$\Delta \overline{G}_i^\circ = \Delta H^\circ - T\Delta S^\circ \tag{III-124}$$

For example, consider reaction (110). Once the enthalpy of formation of each species in the reaction has been obtained these quantities are combined to obtain ΔH° for the reaction according to

$$\Delta H^\circ = \Delta H_f^\circ(H_2) + \Delta H_f^\circ(Hg_2Cl_2) - 2\Delta H_f^\circ(Hg) - 2\Delta H_f^\circ(HCl) \tag{III-125}$$

(the heats of formation of liquid mercury and gaseous hydrogen are zero), and ΔS° is obtained according to

$$\Delta S^\circ = S^\circ(H_2) + S^\circ(Hg_2Cl_2) - 2\,S^\circ(Hg) - 2\,S^\circ(HCl) \tag{III-126}$$

The change in standard Gibbs free energy or chemical potential for the reaction can then be obtained using equation (III-124), and the equilibrium constant can be obtained using equation (III-123).

---

* We shall treat the concept of absolute entropy in Chapter V.

SUMMARY

In this chapter we have introduced the Gibbs free energy, the molar Gibbs free energy and the chemical potential (sometimes called the **partial molar** Gibbs free energy). We obtained the general result that if a system is at equilibrium with respect to a given reaction, then the Gibbs free energy change accompanying that reaction is zero. This result led to the equilibrium constant expressions for reactions of all kinds. The portion of our treatment that preceded the definition and use of the chemical potential, and was therefore in terms of the Gibbs free energy change, can now be interpreted in terms of the more general change in chemical potential. In particular, application of the criterion that the change in chemical potential for a reaction at equilibrium is zero can be used as a starting point to obtain the Clapeyron equation and also the equilibrium constant for ideal gas reactions, both of which were derived in this chapter by a consideration of the Gibbs free energy change.

The important equations developed in this chapter are

(A) *One Component, Phase Equilibria.*

$$dP/dT = \Delta H/T\,\Delta V, \text{ the Clapeyron equation.}$$

$$\ln P_2/P_1 = \frac{\Delta H\,\Delta T}{RT_1T_2}, \text{ the integrated Clausius-Clapeyron equation.}$$

$$\frac{d(\ln P)}{d(1/T)} = -\Delta H/R, \text{ the Clausius-Clapeyron equation.}$$

(B) *Many Component, Many Phase Equilibria.*

$$\Delta\overline{G}_i = \Delta\overline{G}_i^\circ + RT\ln\frac{a_L^{\,l}\,a_M^{\,m}\cdots}{a_B^{\,b}\,a_C^{\,c}\cdots} \quad \text{B, C, \ldots, L, M not at equilibrium concentrations}$$

$$\Delta G_i^\circ = -RT\ln\frac{a_L^{\,l}\,a_M^{\,m}\cdots}{a_B^{\,b}\,a_C^{\,c}\cdots} = -RT\ln K \text{ at equilibrium,}$$

where:

$a_i = P_i$, for ideal gases.

$a_i = \gamma_i m_i$, for solute, practical description.

$a_i = f_i N_i$, for solute or solvent; rational description.

$a_{B_bC_c} = \gamma_\pm^{\,b+c}\, m_\pm^{\,(b+c)}$ for a strong electrolyte in solution.

$a_i = 1$, for a pure condensed substance.

Problems

III-1 From the vapor pressure *versus* temperature data given below, calculate the heat of vaporization of tungsten at 2800°K.

T	2600°K	2700°K	2800°K	2900°K	3000°K
$\log P$	−9.159	−8.528	−7.939	−7.396	−6.886

III-2 There are two allotropic forms of tin, called grey and white tin. The enthalpy change at 298° for the reaction

$$\text{Sn(grey)} \longrightarrow \text{Sn(white)}$$

is 0.50 kcal/mole. The entropy change is 1.745 cal/deg mole. Using the Gibbs free energy criterion for equilibrium and assuming ΔH and ΔS to be independent of T, calculate the transition temperature, i.e., the temperature at which grey and white tin co-exist in equilibrium.

III-3 The standard heat of formation of HgO(s) at 298°K is −21.56 kcal. The standard entropies of HgO(s), Hg(l) and O_2(g) at 298°K are 17.5 cal/deg mole, 18.5 cal/deg mole and 49.0 cal/deg mole. Assuming that ΔH° and ΔS° are independent of T, calculate the temperature at which HgO(s) dissociates into Hg(l) and O_2(g) *in air*.

III-4 The standard enthalpy of formation of ozone at 298°K is 34.0 kcal. The entropies of ozone and oxygen at 298°K are 56.8 cal/deg mole and 49.0 cal/deg mole, respectively. Calculate the equilibrium concentration of ozone in air at 298°K.

III-5 Using the room temperature data tabulated below calculate the maximum work per mole of Na consumed (other than pressure-volume work) that could be obtained at 298°K from the reaction

$$\text{Na(s)} + \text{H}_2\text{O(l)} \longrightarrow \text{NaOH}(\text{m} = 0.10) + \tfrac{1}{2}\text{H}_2\text{(g)}$$

assuming that 0.10 molal NaOH solution behaves ideally.

Substance	ΔH_f°	S°
Na(s)	0	12.2 cal/mole deg
H_2O(l)	−68.3 kcal	16.7 cal/mole deg
NaOH(soln.)	112.2 kcal	11.9 cal/mole deg
H_2(g)	0	31.2 cal/mole deg

III-6 Using the room temperature data tabulated below, calculate the heat of neutralization for H_2O, i.e., the standard enthalpy change for the reaction

$$\text{H}^+\text{(soln.)} + \text{OH}^-\text{(soln.)} \longrightarrow \text{H}_2\text{O(l)}$$

Substance	ΔH_f°
H^+ (soln.)	0
OH^- (soln.)	−54.96 kcal
$H_2O(l)$	−57.80 kcal

III-7 (a) Using the thermodynamic data evaluated at 2000°K and tabulated below, determine whether Be(l) will reduce $Al_2O_3(s)$ at 2000°K. (b) Is there any temperature at which Al(l) will reduce BeO(s)? Assume that ΔH and ΔS are independent of temperature. (c) Which is more important in this reaction, ΔH or $T\Delta S$?

Substance	ΔH_f°(2000°K)	ΔS°(2000°K)
Be(l)	0	15.46 cal/deg mole
BeO(s)	−144.38 kcal	23.47 cal/deg mole
Al(l)	0	22.86 cal/deg mole
$Al_2O_3(s)$	−400.08 kcal	64.80 cal/deg mole

III-8 The vapor pressure of liquid ^4He (the isotope of helium with atomic weight four which boils at 4.2°K) is given as a function of temperature below.

T°K	4.21	4.02	3.81	3.60
PmmHg	760	630	510	402

The standard heat content of copper at room temperature relative to 4.2°K is $H^\circ_{298} - H^\circ_{4.2} = 1{,}195$ cal/mole. How many ml. of liquid ^4He at its boiling point would be evaporated if one gram of copper initially at 298°K were dropped into the ^4He? The density of ^4He(liq) is 0.125g/ml.

CHAPTER IV

Colligative Properties

An incident indelibly imprinted in the mind of one of the authors of this text (B.C.G.) occurred when he was a boy in upper New York State. On a particularly crystal clear cold winter day after a rather bad sleet storm, the porch steps on the house in which he lived were glazed with a thin sheet of ice. His mother called him to the front door, and after sprinkling a plentiful supply of table salt on the steps, said mysteriously, "Watch what happens." With a wonder akin to that experienced by Thoreau when coming upon phosphorescent wood in his trip to the Maine woods, he watched the ice slowly pit where the salt had fallen on it, and then melt down to the bare concrete below.

Few third grade students are so unsophisticated today that they could not have predicted the results of such an experiment. The reason for, and the quantitative description of, the melting process, however, are as obscure to the average college freshman as to the third grader. The phenomenon of the lowering of the melting point of a pure compound in equilibrium with *a solution* of the compound containing an impurity, is one of a trio of properties called **colligative properties.** These properties are so named because they are bound together (colligated) by the fact that they depend only on the number of impurity (solute) particles in solution. The other two of the trio are boiling-point elevation and osmotic pressure.

Thermodynamics provides us with a quantitative description of such behavior. For purposes of admiration, and because the formulae are so simple as to do no harm in exhibiting themselves before uninitiated eyes, we will place them in plain view and discuss them a bit before deriving them.

BOILING-POINT ELEVATION

When a sufficiently small amount of a *nonvolatile* solute (such as sucrose) is added to a solvent (such as water), the vapor pressure of the solution is decreased from that characteristic of the pure solvent in accordance with *Raoult's Law:*

$$P_{\text{solvent}} = P^{\circ}_{\text{solvent}} \cdot \text{N}_{\text{solvent}} \tag{IV-1}$$

where $P^{\circ}_{\text{solvent}}$ is the vapor pressure of the pure solvent and $\text{N}_{\text{solvent}}$ is the mole fraction of the solvent at the temperature in question. Because of this vapor pressure lowering, the solution boils at a higher temperature than the pure solvent. This is to say that since the vapor pressure of the solution *at a given temperature* is lower than that of the pure solvent, it is necessary to heat the solution to a temperature higher than that of the boiling point of the pure solvent to raise the vapor pressure of the solution to atmospheric pressure, i.e. to the boiling point. The relation between the mole fraction of nondissociated solute (such as sugar, not sodium chloride) and the *boiling-point elevation*, ΔT_b, is given approximately by

$$\Delta T_b \cong (RT_b{}^2N_2)/\Delta\overline{H}_{\text{vap}} \tag{IV-2}$$

R is the gas constant, T_b is the boiling point of the pure solvent, $\Delta\overline{H}_{\text{vap}}$ is the molar heat of vaporization of the pure solvent at its normal boiling point, and N_2 is the mole fraction of the solute. If R is expressed in units of (calories/mole deg) and $\Delta\overline{H}_{\text{vap}}$ is expressed in units of calories per mole, then one easily sees that, dimensionally, the equation reads

$$\begin{aligned}\text{boiling-point elevation} &= (\text{cal/mole deg}) \times (\text{deg})^2 \times (\text{mole/cal}) \\ &= \text{deg}\end{aligned}$$

The relation becomes more exact the closer N_2 is to zero, i.e., the less concentrated the solution. Note particularly that except for N_2, all of the terms on the right hand side of equation (IV-2) are constants. Therefore, for a given solvent, the relation may be simply written

$$\Delta T_b \cong K'_b N_2 \tag{IV-3}$$

where K'_b is a *boiling-point constant* for a given solvent. Alternatively, if the mole fraction of the solute is converted to molality by the conversion factor

$$m_2 \cong \frac{1000}{M_1} N_2$$

where M_1 is the molecular weight of the solvent, then the boiling-point elevation expression may be adjusted to read simply

$$\Delta T_b \cong K_b m_2 \qquad \textbf{(IV-4)}$$

where

$$K_b = \frac{M_1}{1000} K'_b = \frac{M_1}{1000} \frac{RT_b^2}{\Delta \overline{H}_{vap}} \qquad \textbf{(IV-5)}$$

With a molar heat of vaporization of 9.7 kcal, a boiling point of 373°K, and a molecular weight of 18 g/mole, for example, the boiling-point constant of water is $0.514 = K_b$, and, therefore, a 0.1 molal solution of sugar would be expected to boil at $(0.514)(0.1) = 0.0514$ degrees above its normal boiling point. Needless to say, this is not a terribly spectacular effect for water, but it becomes more so for solvents with high molecular weights and low heats of vaporization.

If the solute is dissociated into ν species in solution, such as an electrolyte like $Al_2(SO_4)_3$ (here, $\nu = 3 + 2 = 5$), then the boiling-point elevation expression reads

$$\Delta T_b \cong \nu K_b m_2 \qquad \textbf{(IV-6)}$$

Thus, a 0.1 m solution of $Al_2(SO_4)_3$ in water would be expected to boil at 0.256 degrees above the boiling point of water.

It should be noted that the simple measurement of a temperature increment in boiling upon addition of a known weight of undissociated solute yields the molecular weight of the solute, since the boiling-point elevation gives the molality which in turn yields the number of moles of solute, and the molecular weight is simply the ratio of the weight to the number of moles, i.e., $M_2 = w_2/n_2$. For example, if 10 g of a solute (which was determined to be undissociated by, e.g., conductivity measurements) of unknown molecular weight raised the boiling point of 1000 g of water by 0.041 degrees Celsius, the molality, m_2 of the solute would be, by equation (IV-4),

$$m_2 = \Delta T_b / K_b = 0.041/0.514 = 0.0878$$

The molecular weight in this case would be the number of grams per mole, or (10 g/0.0878 moles) = 114 g/mole.

FREEZING-POINT LOWERING

As implied in the introduction to this chapter, ice in equilibrium with a solution of water and sodium chloride melts at a temperature lower than the melting point of pure water. A general statement of this phenomenon is as follows: If a solid and its melt are in equilibrium, and a solute which dissolves in the melt but not in the solid is added, the freezing point of the mixture is lowered. The relation between the *freezing-point lowering*, ΔT_f, and the mole fraction of solute in the melt, N_2, is approximately given by

$$\Delta T_f \cong (N_2 R T_f^2 / \Delta \overline{H}_f) \qquad \textbf{(IV-7)}$$

This formula is exactly analogous to the boiling-point elevation relation, but now T_f is the freezing point of the pure solvent, and $\Delta \overline{H}_f$ is the molar heat of fusion of the pure solvent. Again, using the approximation (IV-3), we can write equation (IV-7) as

$$\Delta T_f \cong K_f m_2 \qquad \textbf{(IV-8)}$$

where m_2 is the molality of the solute in the melt, and K_f, the *freezing-point constant* for a given solvent, is equal to $(R T_f^2 M_1 / 1000\, \Delta \overline{H}_f)$. The freezing-point constant for water is 1.86°C. Therefore a one molal aqueous solution of sucrose would freeze at −1.86°C if the solution were ideal. Again, as with the boiling-point elevation expression, if the solute dissociates into ν particles, the freezing-point lowering is given approximately by the expression

$$\Delta T_f \cong \nu K_f m_2 \qquad \textbf{(IV-9)}$$

For example, ν is 5 for $[Pt(NH_3)_6]\,Cl_4$, and ν is 3 for $BaCl_2$.

Again, we note that the molecular weight of the solute can be determined from its effect upon the freezing point of the solvent provided that no appreciable solid solution forms between solvent and solute (H_2O and NaCl would be an example of a system in which no solid solution formed; D_2O and H_2O would be a case in which one would expect solid solution over the entire composition range. We discuss graphical representations of such systems in Chapter VII). We also note that measurement of freezing-point lowering is in general a much more sensitive method of molecular weight determination than measurement of boiling-point elevation, although both are classical methods of such a determination. It is interesting to realize that zone refining, a process upon which the electronics industry depends for ultra pure semiconductors, is related to the freezing-point lowering phenomenon, as impurities concentrate in the melt if there is a lowering of the freezing point upon the addition of impurity.

OSMOTIC PRESSURE

If a balloon made of ordinary cellophane is filled with an aqueous sugar solution, sealed, and immersed in a container of pure water, it will be observed to swell, and under the proper conditions, finally break. This phenomenon is a result of **osmotic pressure.** The phenomenon of osmosis is related to maintenance of cell fluid solute concentrations and to a certain extent to the movement of sap in trees.

To describe the effect simply, we consider the experiment shown in Fig. 4.1. The solution on the sugar-rich side has risen a height h

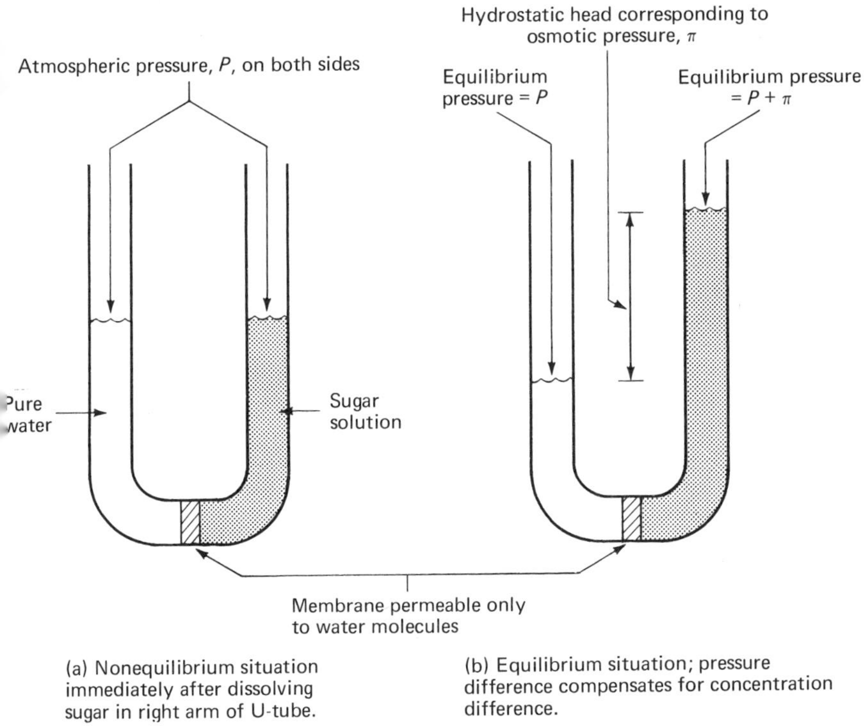

Figure 4.1 Diagram of apparatus for osmotic pressure experiment.

above the level of pure water, producing a hydrostatic head, π, on the right side. The cause of the rise is the tendency for water to pass through the membrane from the compartment containing pure solvent into the one containing the sugar solution. The equation approximately relating the osmotic pressure, π, to the concentration of the solute, N_2, is

$$\pi \overline{V}_1^{\circ} \cong N_2RT \qquad \textbf{(IV-10)}$$

$\overline{V}_1^{\circ}$ is the molar volume of the pure solvent (18 ml/mole in the case of water), and N_2 is the mole fraction of the solute. For sufficiently dilute solutions,

$$N_2 = n_2/(n_1 + n_2) \cong n_2/n_1$$

Therefore, $N_2/\bar{V}_1^\circ = n_2/n_1\bar{V}_1^\circ = n_2/V_1 = C_2$, the molarity of the solute.* Equation (IV-10) may therefore be rearranged to read

$$\pi \cong C_2RT \tag{IV-11}$$

for solutions containing an undissociated solute. For solutions in which the solute dissociates into ν ionic species, (IV-11) reads

$$\pi \cong \nu C_2RT \tag{IV-12}$$

The osmotic pressure is rather spectacularly sensitive to the concentration of the solute. In cases where measurement is practical, therefore, it is a useful measure of solute molecular weight, since knowledge of the concentration of the solution for a known weight of solute added is sufficient to determine the molecular weight.

For example, a 0.1 molar aqueous solution of NaCl, corresponding to 0.078 gram of NaCl per ml of solution, yields an osmotic pressure of

$$\begin{aligned}\pi &= 0.1RT = 2 \times 0.1 \text{ (moles/liter)} \\ &\quad \times 0.082 \text{ (liter-atm/mole-deg)} \times 298 \text{ deg} \\ &= 4.9 \text{ atm}\end{aligned}$$

This pressure corresponds to a column of water forty-nine meters high! This seemingly surprising result is understood in terms of the chemical potential being a sharp function of concentration, but a shallow function of pressure. A rather large (in our terms of reference) pressure difference is therefore required to obtain equilibrium with respect to transfer of a relatively "small" amount of solvent through the membrane.

It is interesting and informative to compare the magnitudes of chemical potential differences due to concentration changes as opposed to pressure changes. The difference due to a concentration change for, say, a species obeying Henry's law is obtained from equation (III-70).

$$\Delta\bar{G} = RT \ln (N_2/N_1)$$

For a solute concentration change from, say, a mole fraction of 0.0001 to a mole fraction of 0.001, the above relation yields a change in

* In stating the second equality, we are using the fact that the moles of solvent times the volume of solvent per mole is equal to the total volume of solvent, i.e. $n_1\bar{V}_1^\circ = V_1$.

chemical potential of

$$\Delta\overline{G} = 2\ (\text{cal/mole-deg}) \times 298\ \text{deg} \times \ln 10$$
$$= 1{,}370\ \text{cal}$$

To obtain the difference due to a pressure change, we recall equation (III-9),

$$dG = -S\,dT + V\,dP$$

which for a pure component may be written in terms of the chemical potential simply by referring to molar quantities;*

$$d\overline{G} = -\overline{S}\,dT + \overline{V}\,dP \qquad \textbf{(IV-13)}$$

from which we obtain

$$\partial\overline{G}/\partial P)_T = \overline{V} \qquad \textbf{(IV-14)}$$

or

$$\overline{G}(P_2) - \overline{G}(P_1) = \int_{P_1}^{P_2} \overline{V}\,dP \qquad \textbf{(IV-15)}$$

Ignoring the pressure dependence of $\overline{V}$ compared to the change in pressure itself (to do so is an article of faith, we know, but you must believe that we really have seen the numbers and know that such an approximation involves no error for present purposes)

$$\Delta\overline{G} = \overline{G}(P_2) - \overline{G}(P_1) = \overline{V}\,\Delta P \qquad \textbf{(IV-16)}$$

To compare the effect of a ten-fold concentration change with a similar pressure change, we consider $\Delta\overline{G}$ due to a pressure change from one to ten atmospheres on water, with a molar volume of 18 ml/mole. By (IV-16)

$$\Delta\overline{G} = \overline{V}\,\Delta P = 18\ (\text{ml/mole}) \times [2\ \text{cal/(mole-deg)}]$$
$$\times\ 1/82\ (\text{ml-atm/mole-deg}) \times 9\ \text{atm}$$
$$= 3.96\ \text{calories}$$

where the conversion factor 2 cal/(mole-deg) ≡ 82 (ml-atm)/mole-deg) has been used to convert from ml-atm to calories. Clearly an enormous pressure change (in terms of what we think of as "ordinary pressures") is needed to yield the same Gibbs Free Energy change as a relatively small concentration change!

---

* It is also the case that for the *i*th component in a solution,

$$d\overline{G}_i = -\overline{S}_i dT + \overline{V}_i\,dP$$

where $\overline{S}_i$ and $\overline{V}_i$ are partial molar entropies and volumes respectively.

DERIVATION OF THE COLLIGATIVE PROPERTY FORMULAE

Boiling-Point Elevation

When a nonvolatile solute is added to a solvent, the vapor pressure of the solvent is lowered by an amount $\overline{P}^\circ - \overline{P}^\circ N_1$, where N_1 is the mole fraction of solvent, providing the solvent obeys Raoult's Law in the concentration range considered. In terms of the concentration of the solute, the vapor pressure lowering of the solvent is

$$\overline{P}^\circ - \overline{P}^\circ(1 - N_2) = \overline{P}^\circ N_2 = \Delta P \qquad \text{(IV-17)}$$

We know approximately the ratio of pressure change to the temperature change along the equilibrium line between vapor and liquid in the concentration range where this expression holds, (generally in very dilute solutions). This change is given by the Clausius-Clapeyron equation:

$$\Delta P/\Delta T \cong dP/dT = \frac{\Delta H \cdot P}{RT^2}$$

Solving for ΔT, utilizing equation (IV-17), and realizing that along the equilibrium line $\overline{P}^\circ$ is P and T is the boiling point of the solution, we find

$$\Delta T \cong \frac{N_2 RT_b{}^2}{\Delta \overline{H}_{vap}} \qquad \text{(IV-18)}$$

For temperatures not far from the boiling point of the pure solvent, the difference between T^2 and the square of the boiling point of the pure solvent is negligible, and we take T in (IV-18) to be T_b, the boiling point of the pure solvent.

A more general derivation given in Appendix IV yields the more exact result

$$\Delta T_b = -\frac{RT_bT \ln a_1}{\Delta \overline{H}_{vap}} \qquad \text{(IV-19)}$$

where a_1 is the activity of the solvent in the liquid and the temperature dependence of the heat of vaporization has been ignored. As discussed in Chapter III, the activity of an undissociated species is given by $a_i = \gamma_i m_i$ or $a_i = f_i N_i$, and the activity of a dissociated species is $\gamma_\pm{}^\nu m_\pm{}^\nu$, where ν is the number of particles into which the species dissociates. The colligative properties are, in fact, among the methods used to determine solvent activities. The activity of the solute in a two component system may be obtained from the activity of the solvent, but it is beyond the purpose of this text to discuss this relation.

It should be particularly noted that equation (IV-19) reduces to

$$\Delta T_b = -\frac{RT_bT}{\Delta \overline{H}_{vap}} \ln N_1 \qquad \text{(IV-20)}$$

for ideal solutions ($f = 1$). While equations (IV-18), (IV-19), and (IV-20), are all "boiling-point elevation" formulae, equation (IV-18) is more of an approximation than equation (IV-19). Equation (IV-18) works well in very dilute solution, provided the activity coefficient is essentially unity; equation (IV-19) is appropriate for the behavior of a solution where N_2 is not $\ll 1$ (see Problem IV-1 at the end of this chapter).

Freezing-Point Lowering

The temperature of a pure solid, say ice, in equilibrium with its melt containing a small amount of a nonvolatile impurity, say sugar in water, is lower than the melting point of the pure material. That this is true is clear from a consideration of the condition for equilibrium; namely the vapor pressure of the solid must equal the vapor pressure of the solvent. If such were not the case in a closed system, either the solid or the solvent would spontaneously vaporize and re-condense as solid or solvent, respectively. If the pure solid is to have a vapor pressure equal to the partial pressure of the solvent in the solution which is *lower* than that of pure solvent by an amount $P°N_2$ in the ideal case (as seen in the boiling-point elevation argument), the solid must necessarily cool. Analytically, at equilibrium,

$$P_{\text{solid}} = P_{\text{solvent in solution}} \qquad \text{(IV-21)}$$

Denoting the pressure at temperature T by the symbol $P(T)$, and the freezing point of the pure solvent by T_f, equation (IV-21) reads

$$P^{\circ\,\text{solid}}(T_f - \Delta T) = P^{\text{solvent}}(T_f - \Delta T) \qquad \text{(IV-22)}$$

The superscript (°) refers to pure component. Equation (IV-22) says that the vapor pressure of the solvent above the solution equals the vapor pressure of the pure solid, *both solution and solid being at the temperature* $T = T_f - \Delta T$. In terms of the vapor pressure lowering of the solvent associated with both cooling and with addition of solute, we write

$$P°(T_f)^{\text{solid}} - \Delta P^{\circ\,\text{solid}} = P°(T_f, N_1 = 1)^{\text{solvent}} - \Delta P^{\text{solvent}} \qquad \text{(IV-23)}$$

Since $P^\circ(T_f)^{\text{solid}} = P^\circ(T_f, N_1 = 1)^{\text{solvent}}$, it follows that

$$\Delta P^{\circ\,\text{solid}} = \Delta P^{\text{solvent}} \tag{IV-24}$$

$\Delta P^{\circ\,\text{solid}}$, the change in vapor pressure with change in temperature of the pure solid, is approximately given by the Clausius-Clapeyron equation, where the enthalpy change involved is the heat of sublimation. For a finite change, we write

$$\Delta P^{\circ\,\text{solid}} = (\Delta \overline{H}_{\text{subl}}\, \Delta T/RT^2) P^\circ(T)^{\text{solid}} \tag{IV-25}$$

The lowering of the vapor pressure of the solvent in the solution, $\Delta P^{\text{solvent}}$, is due to both temperature *and* concentration lowering. We may write

$$\text{solute } (T_f, N_2 = 0) \xrightarrow{\text{Step 1}} \text{solute } (T, N_2 = 0) \xrightarrow{\text{Step 2}} \text{solute } (T, N_2)$$

where it is clear that

$$\Delta P^{\text{solvent}} = \Delta P_1^{\text{solvent}} + \Delta P_2^{\text{solvent}} \tag{IV-26}$$

Clearly,

$$\Delta P_1^{\text{solvent}} \cong \frac{\Delta \overline{H}_{\text{vap}}\, \Delta T}{RT^2} P^{\circ\,\text{solvent}} \tag{IV-27}$$

and

$$\begin{aligned} \Delta P_2^{\text{solvent}} &= P^\circ(T)^{\text{solvent}} - P^\circ(T)^{\text{solvent}} N_1 \\ &= P^\circ(T)^{\text{solvent}} N_2 \end{aligned} \tag{IV-28}$$

Therefore,

$$\Delta P^{\circ\,\text{solid}} = \Delta P^{\text{solvent}} = \Delta P_1^{\text{solvent}} + \Delta P_2^{\text{solvent}}$$

yields

$$\frac{\Delta \overline{H}_{\text{subl}}\, \Delta T}{RT^2} P^\circ(T) = \frac{\Delta \overline{H}_{\text{vap}}\, \Delta T P^\circ(T)}{RT^2} + P^\circ(T) N_2 \tag{IV-29}$$

Rearranging equation (IV-29) and realizing that by the first law, $\Delta \overline{H}_{\text{subl}} - \Delta \overline{H}_{\text{vap}} = \Delta H_{\text{fusion}}$, we obtain

$$\frac{\Delta \overline{H}_{\text{fusion}}\, \Delta T}{RT_f^2} \cong N_2 \tag{IV-30}$$

the freezing-point lowering expression. In equation (IV-30) we have neglected the difference between $1/T^2$ and $1/T_f^2$ for ΔT small compared to T, the usual case. A more elegant derivation in terms of chemical potentials is given in Appendix IV, in which the more exact relation

$$\Delta T_f = -\frac{RT_f T}{\Delta \overline{H}_f} \ln a_1 \tag{IV-31}$$

is derived. The same remarks apply to this relationship as apply to the analogous boiling-point elevation formulae; freezing point depression furnishes one means of activity determination. For solutions in which the solvent obeys Raoult's Law, equation (IV-31) reduces to

$$\Delta T_f = -\frac{RTT_f \ln N_1}{\Delta H_f'} \qquad \text{(IV-32)}$$

This relation is appropriate to the case in which N_2 is not $\ll 1$, again provided the solvent obeys Raoult's law.

Osmotic Pressure

We consider a solution containing solvent at activity a_1, and at imposed pressure $P + \pi$ in equilibrium with pure solvent at unit activity and imposed pressure P (as shown in Fig. 4.1b). The condition for chemical equilibrium is

$$\overline{G}_1^\circ \text{ (at } P) = \overline{G}_1 \text{ (at } P + \pi \text{ and } N_1) \qquad \text{(IV-33)}$$

Writing equation (IV-33) in terms of the activity of the solvent, we have

$$\overline{G}_1^\circ \text{ (at } P) = \overline{G}_1^\circ \text{ (at } P) + RT \ln a_1 \text{ (at } P) + \Delta\overline{G}_1(P \text{ to } P + \pi) \qquad \text{(IV-34)}$$

Expressing the finite change in $\overline{G}_1$ with pressure, $\Delta\overline{G}_1$, as an integral, we have

$$-RT \ln a_1 \text{ (at } P) = \Delta\overline{G}_1(P \text{ to } P + \pi) = \int_P^{P+\pi} \left(\frac{\partial G_1}{\partial P}\right)_T dP$$

$$= \int_P^{P+\pi} \left[\left(\frac{\partial \overline{G}_1^\circ}{\partial P}\right)_T + RT\left(\frac{\partial \ln a_1}{\partial P}\right)_T\right] dP \qquad \text{(IV-35)}$$

By equation (IV-13), $\left(\frac{\partial \overline{G}_1^\circ}{\partial P}\right)_T = \overline{V}_1^\circ$, the molar volume of the pure solvent. If we neglect the pressure dependence of $\overline{V}_1^\circ$ (and it is a good approximation under ordinary conditions to do so) we see that equation (IV-35) reduces to

$$-RT \ln a_1 \text{ (at } P) \cong \overline{V}_1^\circ \int_P^{P+\pi} dP + RT \int_P^{P+\pi} d(\ln a_1)$$

$$\cong \pi\overline{V}_1^\circ + RT \ln a_1 \text{ (at } P + \pi) - RT \ln a_1 \text{ (at } P)$$

or

$$\pi\overline{V}_1^\circ \cong -RT \ln a_1 \qquad \text{(IV-36)}$$

For ideal solutions, $a_1 = N_1$, and (IV-36) reduces to

$$\pi\overline{V}_1^\circ \cong -RT \ln N_1 \qquad \text{(IV-37)}$$

For N_1 sufficiently close to unity, $\ln N_1 = \ln(1 - N_2) \cong -N_2$. (The reader may verify this for himself by comparing $\ln(1 - x)$ with $-x$ for $x = 0.1, 0.01, 0.001$, etc.) Therefore, equation (IV-37) in the limit of very dilute solutions reduces to

$$\pi \overline{V}_1^\circ \cong N_2 RT \qquad \textbf{(IV-38)}$$

or

$$\pi = \frac{N_2}{\overline{V}_1^\circ} RT \simeq \frac{n_2}{n_1 \overline{V}_1^\circ} Rt = C_2 RT \qquad \textbf{(IV-39)}$$

SUMMARY

In this chapter, we have developed the thermodynamics of the colligative properties. In the limit of dilute solutions it was found that boiling-point elevation, freezing-point lowering, and osmotic pressure were approximately proportional to the number of solute particles in solution. The freezing-point lowering and boiling-point elevation formulae in this limit were found to be

$$\Delta T_{f,b} \cong \frac{RT_{f,b}^2 N_2}{\Delta \overline{H}_{f,b}} \cong m_2 K_{f,b}$$

where the subscript f, b is taken to mean fusion or boiling, whichever is appropriate. For concentrated solutions, the appropriate expressions are

$$\Delta T_{f,b} = \frac{RT_{f,b} T}{\Delta \overline{H}_{f,b}} \ln a_1 \cong \frac{RT_{f,b} T}{\Delta \overline{H}_{f,b}} \ln N_1$$

The osmotic pressure was given by

$$\pi \cong \frac{N_2}{\overline{V}_1^\circ} RT \cong C_2 RT$$

for dilute solutions, and by

$$\pi = \frac{RT}{\overline{V}_1^\circ} \ln a_1 \cong \frac{RT}{\overline{V}_1^\circ} \ln N_1$$

for solutions where N_2 is not much less than unity.

Caveat Regarding the use of Approximate Formulae

In this chapter, we have tried to make clear the "approximately equal" sign ($\cong$) that some formulae we developed are approximate. This sign is to be taken seriously if ridiculous physical answers are not to be obtained from a calculation. Problem 1 in this chapter is a case in point. If one unthinkingly uses equation (IV-7) for this calculation, he finds a value of N_2 as shown below

$$\frac{\Delta\overline{H}_{fus}\,\Delta T}{RT_f^2} = \frac{1440 \times 29}{1.986 \times (273)^2} = 0.28 = N_2$$

If one believed this number and filled his radiator accordingly, he would find in all probability a plugged thermostat or a cracked engine block at temperature not too much lower than −15°C!

Problems

IV-1 Ethylene glycol, $CH_2(OH)CH_2(OH)$, is a commonly used anti-freeze for car radiators. How many liters of this compound (of density 1.11) are required for a 20 liter car radiator to insure that the radiator contents will not freeze above −20°F (−29°C). $\Delta\overline{H}_f(H_2O) = 1440$ cal/mole.

IV-2 Benzene (C_6H_6) freezes at −69°C. The freezing point of a solution 0.01 mole of aniline in 100 g of benzene is 0.49 degrees below this temperature. What is the molar heat of fusion of benzene?

IV-3 6.106 grams of an undissociated solute in a liter of water results in an osmotic pressure of 1.2 atm.

(a) What is the molecular weight of the solute?
(b) What would be the boiling point of such a solution?

CHAPTER V

The Third Law and High Temperature Equilibrium Calculations

THE THIRD LAW OF THERMODYNAMICS

It frequently happens in science that advances in one area have significant consequences in seemingly unrelated areas. Such is the case with the third law of thermodynamics. This principle, first enunciated by Nernst and Planck in the early 1900's on the basis of *low temperature* information and confirmed for some specific cases by experiments at *low temperatures* by Eastman and coworkers and Giauque and coworkers, is extremely useful in facilitating calculations at *high temperatures*.

The Third Law of Thermodynamics can be stated in the following manner: *The entropy of all perfect crystalline solids is zero at absolute zero*. There is something of a problem concerning what is meant by a "perfect crystalline solid." We shall define a *perfect crystalline solid* as a solid which is in equilibrium with respect to changes which occur in it without changes of phase, e.g., diffusional changes. A more sophisticated definition results from statistical considerations which are beyond the scope of this text.

The utility of this law lies in the fact that we can calculate the entropy of a substance from a knowledge of its heat capacity as a function of temperature. Since

$$\frac{dq^{\text{rev.}}}{T} = \frac{C_p\,dT}{T} \qquad \text{(V-1)}$$

for a process at constant pressure, we have (if there are no phase changes)

$$S^\circ(T_2) - S^\circ(T_1) = \int_{T_1}^{T_2} \frac{C_p{}^\circ}{T}\, dT \qquad \text{(V-2)}$$

or, if $T_1 = 0$, by the third law

$$S^\circ(T) = \int_0^T \frac{C_p{}^\circ}{T}\, dT \qquad \text{(V-3)}$$

Thus we have expressed the absolute value of the entropy (as contrasted to ΔS) as a function of temperature. If there are phase changes, melting for example, the entropy change for such a process is given by

$$\Delta S^\circ_{\text{fusion}} = \frac{\Delta H^\circ_{\text{fusion}}}{T_{\text{fusion}}} \qquad \text{(V-4)}$$

and therefore

$$S^\circ_{\text{liquid}}(T) = \int_0^{T_f} \frac{C_p{}^\circ(\text{solid})}{T}\, dT + \frac{\Delta H^\circ_{\text{fusion}}}{T_{\text{fusion}}} + \int_{T_f}^{T} \frac{C_p{}^\circ(\text{liquid})}{T}\, dT \qquad \text{(V-5)}$$

THE FREE ENERGY FUNCTION AND CHEMICAL EQUILIBRIUM

The free energy function, abbreviated fef, is a highly useful property, the usefulness of which will be illustrated later in this chapter. It is defined to be the difference between the Gibbs free energy of a substance in its standard state at some temperature T, $G_T{}^\circ$, and the Gibbs free energy of the substance in (usually) its standard state for the state of aggregation in question at 0°K ($G_0{}^\circ = H_0{}^\circ$) divided by the temperature, T. That is,

$$\text{fef} = \frac{G_T{}^\circ - H_0{}^\circ}{T} \qquad \text{(V-6)}$$

The usefulness of the free energy function derives primarily from the fact that the change in the free energy function, Δfef, for a chemical reaction is simply related to the equilibrium constant for the reaction in question, according to

$$\Delta\text{fef} = \Delta G_T{}^\circ/T - \Delta H_0{}^\circ/T = -R \ln K - \Delta H_0{}^\circ/T \qquad \text{(V-7)}$$

In order to better grasp the idea of the fef let us consider a concrete example such as H_2O. In the case of gaseous water, $H_0{}^\circ$ would be the enthalpy of one mole of gaseous water in the hypothetical ideal gas state at one atmosphere pressure and 0°K, and $G_T°$ would be the Gibbs free energy of one mole of gaseous water in the hypothetical ideal gas state at the temperature in question. On the other hand, the fef of solid water would involve the enthalpy of one mole of ice (in the crystalline modification in question), and, strictly speaking, with an applied pressure of one atmosphere, and the Gibbs free energy value, $G_T°$, would be that for the solid at the temperature in question, and, again strictly speaking, at an applied pressure of one atmosphere.

If we recall that

$$G_T° = H_T° - TS_T° \qquad \text{(V-8)}$$

we see that

$$\text{fef} = \frac{H_T° - TS_T° - H_0°}{T} = \frac{H_T° - H_0°}{T} - S_T° \qquad \text{(V-9)}$$

by the definition of G. Consider a solid which undergoes no transitions when heated. We can calculate both terms on the right hand side of equation (V-9) from heat capacity data alone, $S_T°$ by equation (V-3), and $H_T° - H_0°$ by equation (I-45). If the solid undergoes a transition, such as a change in crystal structure or melting, then the appropriate equations to use are equation (V-5) and equation (I-46), which would yield the fef with $H_0°$ corresponding to the enthalpy of the low temperature solid at 0°K. To change this fef so that it will be in accord with the definition given in equation (V-6), a value for $\Delta H_0°$ for the phase transition is required. It is frequently the case that this value must be estimated by the use of *estimated* heat capacities, for only when the high temperature phase can be obtained at low temperatures, i.e., in a metastable state, can the heat capacities for the high temperature phase be obtained experimentally at temperatures below the transition temperature. For example, we can obtain monoclinic sulfur at temperatures below 368.5°K (the rhombic to monoclinic transition temperature) but not liquid water at temperatures much below 273.15°K. The use of estimated heat capacities in no way effects the accuracy of calculations at high temperatures, however, for when we use the fef we also use the value of $\Delta H_f°$ (0°K) *from the same tabulation*, and the same estimated heat capacity is used to calculate both the fef and $\Delta H_f°$. Thus the quantities calculated from the estimated heat capacities cancel exactly, and the purpose of the estimate is only to make the use of the tables a routine matter, i.e., all tables are used in the same way whether they be for liquids, solids, or gases. The significance of these remarks will probably become clear only after some familiarity is gained with the use of the fef's.

For purposes of determining fef's of chemical substances it is usually sufficient to determine the heat capacity and transition enthalpies down to about 1°K. Most phenomena which occur below 1°K do not have chemically significant heat effects associated with them. The heat capacity of a solid at temperatures in this region was shown by P. Debye in 1912 to be proportional to the cube of the temperature. Thus the method of calculating $H_T^\circ - H_0^\circ$ and S_T° at very low temperatures is well known, namely,

$$H_T^\circ - H_0^\circ = \int_0^T a\mathrm{T}^3 \, d\mathrm{T} \tag{V-10}$$

and

$$S_T^\circ = \int_0^T a\mathrm{T}^2 \, d\mathrm{T} \tag{V-11}$$

where the constant a is determined from the heat capacities determined at the lowest experimental temperatures.

We thus have experimental methods and theory which enable the determination of the fef of a substance at any temperature, T, provided the heat capacity and transition enthalpies are known from about 1°K to the temperature, T. The fef's of gaseous substances are usually obtained by calculation from spectroscopic data utilizing equations obtained from statistical considerations, but could *in principle* be obtained by calorimetric measurement as described above. By experiment and theory, fef's have been obtained for many elements and compounds and have been collected in such compilations as the JANAF tables (see Appendix I).

There is one minor practical point that should be mentioned with regard to tabulated fef's, namely the enthalpy reference temperature. In the above discussion and in equation (V-6) this temperature was taken to be 0°K, and so it is in some tabulations. In others it is taken to be 298°K and the fef is defined as

$$\text{fef} = \frac{H_T^\circ - H_{298}^\circ}{\mathrm{T}} - S_T^\circ \tag{V-12}$$

The difference between the two fef's is $\dfrac{H_{298}^\circ - H_0^\circ}{\mathrm{T}}$ mathematically, and in practice is the use of ΔH_0° or ΔH_{298}° in equation (V-7). The significance of the last half of this remark will become apparent, as will the use of the fef's in high temperature calculations, as you read through the three examples presented below.

EXAMPLES OF EQUILIBRIUM CALCULATIONS USING fef's

Example 1. As a first example we consider the breaking of the oxygen-oxygen bond in O_2. We might want to know, for example, how much atomic oxygen and how much molecular oxygen are present at 2,000°K and a total pressure of 10^{-6} atm. We consider the reaction

$$O_2(g) = 2\,O(g) \tag{V-13}$$

For this reaction the difference between the free energy functions of products and those of reactants is

$$\Delta\text{fef} = 2\text{fef}(O) - \text{fef}(O_2) \tag{V-14}$$

and is also

$$\Delta\text{fef} = \frac{\Delta G_T{}^\circ - \Delta H_0{}^\circ}{T} \tag{V-15}$$

if the free energy function relative to 0°K is used. We can evaluate equation (V-14) by looking up the fef's in one of the compilations listed in Appendix I and performing the indicated arithmetic. We obtain

$$\Delta\text{fef} = -29.90\text{ cal/deg} \tag{V-16}$$

at 2,000°K for the reaction as written. We also obtain the standard enthalpies of formation of reactants and products at 0°K from a standard compilation, and obtain $\Delta H_0{}^\circ = 117.96$ kcal for the reaction as written. Combination of the above yields for the Gibbs free energy change for the reaction at 2,000°K

$$\Delta G^\circ = 58.16\text{ kcal} \tag{V-17}$$

We recall that

$$\Delta G_T{}^\circ = -RT\ln K = -2.303RT\log K \tag{V-18}$$

$$= -4.576 \cdot 2{,}000 \cdot \log \frac{P_O{}^2}{P_{O_2}} \tag{V-19}$$

and obtain

$$\log P_O{}^2/P_{O_2} = -\frac{58{,}160}{2{,}000 \times 4.576}\,\frac{\text{cal}}{\text{degrees} \times \text{cal/degree}} \tag{V-20}$$

$$= -6.355$$

Thus

$$P_O{}^2/P_{O_2} = 4.42 \times 10^{-7} \tag{V-21}$$

The total pressure was specified as 10^{-6} atm; thus

$$P_O + P_{O_2} = 10^{-6} \tag{V-22}$$

Combining these equations for P_O and P_{O_2} yields

$$\frac{P_O^2}{10^{-6} - P_O} = 4.42 \times 10^{-7} \quad \text{(V-23)}$$

or

$$P_O^2 + 4.42 \times 10^{-7} P_O - 4.42 \times 10^{-13} = 0 \quad \text{(V-24)}$$

Using the quadratic formula

$$P_O = \frac{-4.42 \times 10^{-7} \pm \sqrt{19.5 \times 10^{-14} + 17.7 \times 10^{-13}}}{2}$$

$$= 4.8 \times 10^{-7} \text{ atm} \quad \text{(V-25)}$$

and hence

$$P_{O_2} = 5.2 \times 10^{-7} \text{ atm} \quad \text{(V-26)}$$

Thus at 2,000°K and 10^{-6} atm total pressure, gaseous oxygen is approximately 48% dissociated into atomic oxygen.

The first example utilized the fef's relative to 0°K and we used the enthalpies of formation at 0°K to obtain $\Delta H_0°$, the enthalpy change appropriate to the calculation. In the second example we shall utilize the fef's relative to 298°K simply to illustrate the differences (which are few), and the similarities, between the two calculations.

Example 2. Let us in this calculation consider the dissociation of a somewhat more interesting gaseous molecule, namely $Al_2O(g)$, a well known species at high temperatures, and the gaseous dissociation reaction

$$Al_2O(g) = AlO(g) + Al(g) \quad \text{(V-27)}$$

In order to obtain results which we can readily compare with the dissociation of O_2 discussed in example 1, we shall again fix the temperature at 2,000°K and the total pressure at 10^{-6} atm. The appropriate difference between the free energy functions for the reaction as written is

$$\Delta\text{fef} = \text{fef}[Al(g)] + \text{fef}[AlO(g)] - \text{fef}[Al_2O(g)] \quad \text{(V-28)}$$

By consulting the tables, we find Δfef $= -30.3$ cal/deg at 2,000°K for the fef's relative to 298°K. The enthalpies of formation of the various gaseous species yield

$$\Delta H^\circ_{298} = 130 \pm 10 \text{ kcal} \quad \text{(V-29)}$$

The rather large uncertainty in this value arises primarily from a large uncertainty in the enthalpy of formation of $AlO(g)$. We obtain the equilibrium constant, K, for the reaction under consideration by substitution of these values for ΔH°_{298} and Δfef into

$$\Delta\text{fef} = -4.576 \log K - \frac{\Delta H^\circ_{298}}{T} \qquad \text{(V-30)}$$

taking care to express ΔH°_{298} in consistent units, calories in this case, rather than kilocalories. We obtain log $K \cong -7.6$ or $K \cong 2.5 \times 10^{-8}$, with an uncertainty of about a factor of ten arising from the uncertainty in the enthalpy change at 298°K.

This value can now be used to determine the degree of dissociation of Al_2O(g) into AlO(g) and Al(g). We have

$$\frac{P_{AlO}P_{Al}}{P_{Al_2O}} = 2.5 \times 10^{-8} \qquad \text{(V-31)}$$

and

$$P_{AlO} = P_{Al} + P_{Al_2O} = 10^{-6} \text{ atm} \qquad \text{(V-32)}$$

By the stoichiometry of the reaction we have

$$P_{Al} = P_{AlO} \qquad \text{(V-33)}$$

Combining these three equations yields

$$\frac{P^2_{AlO}}{10^{-6} - 2P_{AlO}} = 2.5 \times 10^{-8} \qquad \text{(V-34)}$$

If we assume that $2P_{AlO} \ll 10^{-6}$, we can neglect the $2P_{AlO}$ in the denominator and obtain

$$P^2_{AlO} = 2.5 \times 10^{-14} \qquad \text{(V-35)}$$

or

$$P_{AlO} = 1.58 \times 10^{-7} \text{ atm} \qquad \text{(V-36)}$$

We see that our approximation was not quite correct, and recalculate P_{AlO} taking the denominator in equation (V-34) to be

$$10^{-6} - (2 \cdot 1.58 \times 10^{-7}) = 0.68 \times 10^{-6} \qquad \text{(V-37)}$$

We obtain an improved value for the pressure of AlO, namely

$$P_{AlO} = \sqrt{0.68 \times 10^{-6} \cdot 2.5 \times 10^{-8}} = 1.3 \times 10^{-7} \text{ atm} \qquad \text{(V-38)}$$

We could iteratively improve upon this value, but it is not worthwhile to do so, especially considering the uncertainty associated with the equilibrium constant. Our result, then, is that the best value for the degree of dissociation of Al_2O(g) into AlO(g) and Al(g) is 13%, but that there is a high degree of uncertainty associated with this figure.

Example 3. The third example is one of an equilibrium between a condensed phase and a vapor phase. It provides an insight into the complexity of a high temperature vapor in equilibrium with a condensed phase. We have chosen the vaporization of graphite as our example, and shall calculate the partial pressures of C(*g*), $C_2(g)$ and $C_3(g)$ over graphite at 2,000°K. For the reactions

$$\begin{aligned} C(s) &= C(g) \\ 2\,C(s) &= C_2(g) \end{aligned} \qquad \text{(V-39)}$$

and

$$3\,C(s) = C_3(g)$$

we obtain Δfef's at 2,000°K of −36.83, −45.10 and −45.61 cal/deg, respectively, where the fef's are relative to 0°K. The $\Delta H_0°$'s for the reactions are 169.6, 196 and 188 kcal, respectively. Doing the usual calculations we obtain

$$\begin{aligned} K_1 &= P_C = 3.2 \times 10^{-11}\text{ atm} \\ K_2 &= P_{C_2} = 2.5 \times 10^{-12}\text{ atm} \\ K_3 &= P_{C_3} = 2.5 \times 10^{-11}\text{ atm} \end{aligned} \qquad \text{(V-40)}$$

Thus it appears that graphite is rather nonvolatile, and that C(*g*) and $C_3(g)$ are roughly ten times as important as $C_2(g)$ in the vapor over C(*s*) at 2,000°K.

SUMMARY

In this chapter we have briefly discussed the Third Law of Thermodynamics. We have shown that as a result of the third law the equilibrium constant for a chemical reaction can be obtained in terms of a temperature independent quantity $\Delta H_0°$ and a temperature dependent quantity, Δfef. The temperature independent quantity is the difference in the ground state energies of products and reactants, whereas the temperature dependent quantity, Δfef, includes the entropy difference and the effect of variation of enthalpy with temperature, i.e., accounts for all effects that result from population of higher energy levels by both reactants and products. The important equation developed is

$$\Delta\text{fef} = -R \ln K - \frac{\Delta H_0°}{T}$$

Problems

V-1 Calculate the degree of dissociation (α) of nitrogen at 2,000°K and a total pressure of 10^{-6} atm. Use the data tabulated below

	$\Delta H_f°(0°K)$	$\left(\frac{G_T° - H_0°}{T}\right)_{2000}$
$N_2(g)$	0	52.48 cal/mole deg
$N(g)$	112.5 kcal	−41.10 cal/mole deg

V-2 For the reaction

$$H(g) = H^+(g) + e^-(g)$$

the enthalpy change at 0°K is 313.61 kcal. Using the free energy functions for H(g), $H^+(g)$ and e^- at 5,000°K, which are −36.43 cal/deg mole, −35.05 cal/deg mole and −14.02 cal/deg mole, respectively, calculate the partial pressure of H(g), $H^+(g)$ and $e^-(g)$ at 5000°K and a total pressure of 10^{-10} atm.

V-3 From statistical mechanics it is found that the free energy function of an ideal gas can be considered in two parts—translational and internal. It is further found that the translational part is given by

$$-\left(\frac{G° - H_0°}{T}\right)_{\text{trans}} = 6.864 \log M + 11.440 \log T - 7.283 \text{ cal/deg}$$

where M = the molecular weight of the gas and T = the temperature in degrees Kelvin, and that the internal part is given by

$$-\left(\frac{G° - H_0°}{T}\right)_{\text{int}} = 4.576 \log\left(\sum_i g_i e^{\frac{-(\bar{E}_i - \bar{E}_0)}{RT}}\right)$$

where $\bar{E}_i - \bar{E}_o$ is the energy of one mole of substance in the *ith* energy state (the ground state corresponding to i = o), R is the gas constant in units consistent with the energy units used, and g_i is the "degeneracy" of the ith state. Calculate equilibrium constants for the reactions below given the following information.

$$Ca°(g) = Ca^+(g) + e^-(g) \qquad \Delta H_0° = 140 \text{ kcal}$$
$$Ca^+(g) = Ca^{2+}(g) + e^-(g) \qquad \Delta H_0° = 273 \text{ kcal}$$

The ground state of Ca° is singly degenerate ($g_0 = 1$), the first excited state lies 15,357 cm^{-1} above the ground state and is five-fold degenerate ($g_1 = 5$), the second excited state lies 20,371 cm^{-1} above the ground state and is seven-fold degenerate. The ground state of Ca^+ is two-fold degenerate, the first excited state lies 13,710 cm^{-1} above the ground state and is six-fold degenerate. The ground state of Ca^{2+} is singly degenerate and the only state of the free electron is two-fold degenerate. These states are the only ones you need consider. You can assume that the gases are ideal. To convert from cm^{-1} to cal/mole you multiply by 2.859 cal/cm^{-1} mole.

V-4 Calculate the partial pressure of potassium monomer and dimer [K(g) and K_2(g)] at the normal boiling point of potassium (1037°K). The data to be used are tabulated below.

	$\Delta H_f°(0°K)$	$\left(\frac{G° - H_0°}{T}\right)_{1000}$	$\left(\frac{G° - H_0°}{T}\right)_{1100}$
K(g)	21.52 kcal	−39.34 cal/mole deg	−39.82 cal/mole deg
K_2(g)	31.19 kcal	−61.75 cal/mole deg	−62.63 cal/mole deg

V-5 Calculate the partial pressure of CH_2(g) in equilibrium with solid carbon and hydrogen if the total pressure of hydrogen is 10^{-6} atm and the temperature is 2000°K. Use the data tabulated below.

	$\Delta H_f°(0°K)$	$\left(\frac{G° - H_0°}{T}\right)_{2000}$
$CH_2(g)$	95.18 kcal	−52.27 cal/deg mole
$H(g)$	51.63 kcal	−31.88 cal/deg mole
$C(s)$	0	−5.37 cal/deg mole
$H_2(g)$	0	−37.67 cal/deg mole

V-6 Calculate the pressure of Ti(g) in equilibrium with solid titanium metal at 1500°K. Use the data tabulated below.

	$\Delta H_f°(0°K)$	$\left(\frac{G° - H_0°}{T}\right)_{1550}$
$Ti(s)$	0	−12.23 cal/mole deg
$Ti(g)$	111.84 kcal	−46.24

V-7 The vapor pressure of liquid aluminum at 2000°K is $10^{-2.139}$ atm. Using the free energy functions relative to 298°K which are tabulated below, calculate the standard enthalpy of vaporization of aluminum at 298°K. The vapor at 298°K consists only of Al(g).

	$\left(\frac{G° - H_{298}°}{T}\right)_{2000}$
$Al(l)$	−14.823
$Al(g)$	−44.583

CHAPTER VI

Electrochemistry and Chemical Equilibrium

Chemists are interested in electrochemical phenomena for many reasons. In the first place, much of the thermodynamic data which are available for ionic solute species in H_2O at 298°K were obtained by electrochemical studies. Furthermore, many practical devices currently in use, such as the dry cell, the lead storage battery and the nickel-cadmium cell can be understood in thermodynamic terms by considering the application of thermodynamics to electrochemical reactions. The conversion of "chemical energy" or bond energy into electrical energy by means of a fuel cell is a hoped for development which will permit the conversion of energy without requiring the production of heat, and therefore the ultimate limitation of the second law on the conversion efficiency. Thus, this electrochemical device is of interest in the hope of obtaining 100% energy conversion. Finally, it should be mentioned that many life processes are electrochemical in nature. We shall discuss only the simplest of electrochemical phenomena, namely the emf of a reversible galvanic cell. This discussion, since it is restricted to reversible processes, avoids the many challenging questions introduced by the consideration of questions concerning how rapidly or by what path a reaction occurs. It does, however, permit the application of thermodynamics.

AN EXAMPLE

Let us start by considering a solution of hydrochloric acid in which are immersed two electrodes. An electrode is a combination of substances, at least one of which is a conductor of electricity, so arranged that a chemical reaction can occur to produce an electron in the conductor. For example, consider an acid solution into which a piece of platinum metal is dipped, and through which hydrogen gas is bubbled such that it comes in contact with the platinum. This is called the **hydrogen electrode** and the chemical reaction which occurs is

$$\tfrac{1}{2}\,H_2(g, P) = H^+(\text{soln.}) + e^{-*} \qquad \textbf{(VI-1)}$$

Another example of an electrode is a piece of silver metal on which solid AgCl is deposited and which is dipped into a solution containing chloride ions. This electrode is called the silver-silver chloride electrode and the chemical reaction which occurs is

$$Ag(s) + Cl^-(\text{soln.}) = AgCl(s) + e^- \qquad \textbf{(VI-2)}$$

Now let us consider the immersion of both of these unrelated electrodes into the HCl solution, as in Fig. 6.1. In general, there is a tendency for the chemical reaction

$$\tfrac{1}{2}\,H_2(g, P) + AgCl(s) = Ag(s) + H^+(\text{soln.}) + Cl^-(\text{soln.}) \qquad \textbf{(VI-3)}$$

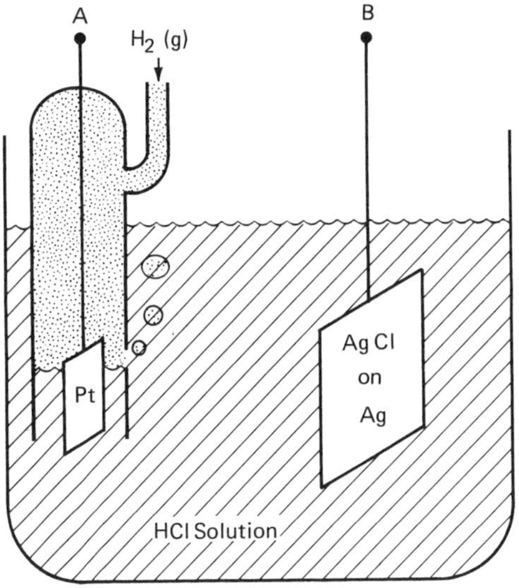

Figure 6.1 Diagram of an electrochemical cell consisting of a hydrogen electrode and a silver-silver chloride electrode.

---

* This reaction can occur in either the forward or reverse direction.

to proceed in either the forward or the reverse direction, depending upon the experimental condition. If a conducting wire were connected from B to A, the tendency for the reaction to proceed in the forward direction would result in a flow of electrons from A to B, which could be observed with an ammeter. If the reaction tended to occur in the reverse direction the electrons would flow from B to A, which could be similarly observed. Thus we could say, by observing an ammeter, whether the particular reaction of hydrogen gas at pressure P with solid silver chloride to form solid silver plus hydrochloric acid at some mean molality $m_{\pm}$ tended to proceed to the left or to the right. The generalization of equation (III-113) to the case of the general chemical reaction yields, as in the special case, that the derivative dG/dn is equal to the change in chemical potential for the reaction, $\Delta\overline{G}_{\mathrm{i}}$. Recall from our discussion of Chapter II that if a process (in this case the chemical reaction) is spontaneous as written, then the change in the Gibbs free energy (in this case dG/dn) is negative, and if the reaction is spontaneous in the reverse direction then dG/dn is positive, with $dG/dn = 0$ corresponding to equilibrium with respect to the reaction. Thus from the direction of spontaneous flow of electrons from one electrode to another we could determine the sign of $\Delta\overline{G}_{\mathrm{i}}$ for the chemical reaction, reaction (VI-3) in the example under discussion.

EMF AND GIBBS FREE ENERGY CHANGES

We can, however, with the aid of a potentiometer do a great deal more than this. A potentiometer is a device which the experimental scientist uses to measure voltages precisely. It provides an accurately calibrated voltage drop across a precisely known length of resistance wire, and a means for picking off the voltage drop across a precisely known fraction of that length, and thus an accurately known, variable voltage, at two terminals. Now let us suppose that we place a potentiometer across the terminals A and B of Fig. 6.1 in such a way as to exactly counterbalance the flowing tendency for the electrons produced and consumed at the electrodes. We would observe this balance by means of a very sensitive current measuring device, a galvanometer. In practice the balance is not exact, but currents as low as 10^{-9} ampere can be detected with modern techniques, and thus a balance very close to zero current can be obtained. We shall assume that the balance is perfect for the rest of this discussion, since the deviation introduces no significant error. When the balance is obtained there is no tendency for electrons to flow from A to B or B to A. Since the driving force for the flow of electrons from A to B is entirely that of the chemical reaction, the counterbalancing voltage provided by the potentiometer is a precise measure of the tendency for the reaction to occur. In fact, the maximum obtainable (non-*PV*) work is the charge carried times this voltage, *if* a current were to flow. Thus the electrical work that *would* be obtained if one mole of electrons flowed through the potentiometer in a reversible manner is just the maximum obtainable (non-*PV*) work from the above reaction if it proceeded so as to produce and consume one mole of electrons, i.e., it is the maximum work obtainable ($-w'$) per equivalent of chemical reaction. We define the *emf* of the cell to be the voltage when no current flows, and then calculate the maximum available work by multiplying this *emf*, E, by the appropriate charge. In general, then

$$-w' = nF\mathrm{E} \qquad \text{(VI-4)}$$

where n is the number of electrons produced and consumed in the *half reactions* [reactions such as (VI-1) and (VI-2)] and F is the charge on one equivalent of electrons. We recall that the maximum available work is equal to the negative of the change in Gibbs free energy and thus

$$dG/dn = -nF\mathrm{E} \qquad \text{(VI-5)}$$

and

$$\Delta\overline{G}_{\mathrm{i}} = -nF\mathrm{E} \qquad \text{(VI-6)}$$

For the reaction considered above we have, from Chapter III

$$\Delta\overline{G}_{\mathrm{i}} = \Delta\overline{G}_{\mathrm{i}}^{\circ} + \mathrm{RT}\ln\frac{\gamma_{\pm}^{2}m_{\pm}^{2}}{P_{\mathrm{H}_2}^{1/2}} \qquad \text{(VI-7)}$$

and combining equations (VI-6) and (VI-7) yields

$$\mathrm{E} = \mathrm{E}^{\circ} - \frac{\mathrm{RT}}{nF}\ln\frac{\gamma_{\pm}^{2}m^{2}}{P_{\mathrm{H}_2}^{1/2}} \qquad \text{(VI-8)}$$

where we have introduced $\Delta\overline{G}_{\mathrm{i}}^{\circ} = -nF\mathrm{E}^{\circ}$.

This equation is called the **Nernst equation.** Note that the definition of E° for a cell is such that it is the *emf* that would be developed by the cell if the reactants and products were in their standard states (standard states were discussed in Chapters I and II). We have thus obtained a result of central importance to the inter-relationship of chemistry and electricity; namely, if the appropriate electrodes can be found and the *emf* measured we can relate this measured *emf* to the activities of the reactants and the products in a fairly simple and straightforward way.

In order to evaluate expressions of the sort obtained above we need to know the numerical value of the Faraday. In electrical units it is 96,490 coulombs per mole, while in chemically more useful units it is 23,060 calories per volt per mole. Using the second value, we obtain for equation (VI-8) at room temperature

$$\mathrm{E} = \mathrm{E}^\circ - \frac{1.987 \text{ cal mole}^{-1}\,{}^\circ\mathrm{K}^{-1} \cdot 298^\circ\mathrm{K} \cdot 2.303}{23{,}060 \text{ cal volt}^{-1} \text{ mole}^{-1}} \log \frac{\gamma_\pm^{\ 2} m_\pm^{\ 2}}{P_{\mathrm{H}_2}^{\ 1/2}} \qquad \textbf{(VI-9)}$$

or

$$\mathrm{E} = \mathrm{E}^\circ - 0.05916 \log \frac{\gamma_\pm^{\ 2} m_\pm^{\ 2}}{P_{\mathrm{H}_2}^{\ 1/2}} \qquad \textbf{(VI-10)}$$

This equation can be used to demonstrate how E° and ΔG° for reaction (VI-3) can be obtained. We can rearrange equation (VI-10) to obtain

$$\mathrm{E} + 0.05916 \log \frac{m_\pm^{\ 2}}{P_{\mathrm{H}_2}^{\ 1/2}} = \mathrm{E}^\circ - 0.05916 \log \gamma_\pm^{\ 2} \qquad \textbf{(VI-11)}$$

All of the quantities on the left can be measured: E as described in the beginning of this chapter, $m_\pm$ by chemical titration, and P_{H_2} with a manometer. If the values of the left hand side of equation (VI-11) are appropriately plotted so as to extrapolate to zero concentration, then

$$\lim_{m_\pm \to 0} \left(\mathrm{E} + 0.05916 \log \frac{m_\pm^{\ 2}}{P_{\mathrm{H}_2}^{\ 1/2}} \right) = \mathrm{E}^\circ \qquad \textbf{(VI-12)}$$

since $\lim_{m_\pm \to 0} \gamma_\pm = 1$ and, of course, $\log 1 = 0$. The appropriate quantity against which to plot the left side of equation (VI-11) is the square root of the ionic strength $\sqrt{\mathrm{I}}$, where the ionic strength is defined as

$$\mathrm{I} = \tfrac{1}{2} \sum m_i Z_i^2 \qquad \textbf{(VI-13)}$$

in which m_i is the molality of the i'th species and Z_i is the charge of the i'th species. In the case of an HCl solution the ionic strength is equal to the molality of the solution, and thus the extrapolation discussed above should be made against $\sqrt{m_{\mathrm{HCl}}}$. The fact that plotting *vs.* $\sqrt{\mathrm{I}}$ yields a linear limiting plot in cases such as those discussed above was first recognized by G. N. Lewis as an empirical fact, and was later justified by P. Debye and E. Hückel in a theory published in 1923.

Of course everything we have developed above for the case of the reaction of hydrogen with silver chloride holds for the general reaction

$$aA + bB + \cdots = lL + mM + \cdots$$

and, provided a suitable pair of electrodes can be found, the *emf* at 298°K is given by

$$E = E^\circ - \frac{0.05916}{n} \log \frac{a_L^{\ l} a_M^{\ m} \cdots}{a_A^{\ a} a_B^{\ b} \cdots} \qquad \textbf{(VI-14)}$$

TEMPERATURE DEPENDENCE OF THE EMF

As we have just seen, we can obtain information about the tendency for reactions to occur, and the magnitude of the Gibbs free energy change for the reaction, from the *emf* for the reaction. We can also obtain the standard entropy and enthalpy changes for the reaction by investigating the temperature dependence of the *emf.* From equation (III-9) it is obvious that

$$\left.\frac{\partial \Delta G}{\partial T}\right)_P = -\Delta S \qquad \textbf{(VI-15)}$$

If we substitute for ΔG according to equation (VI-6), we obtain

$$\left.\frac{\partial E}{\partial T}\right)_P = \frac{\Delta S}{nF} \qquad \textbf{(VI-16)}$$

and from the temperature dependence of E°

$$\left.\frac{\partial E^\circ}{\partial T}\right)_P = \frac{\Delta S^\circ}{nF} \qquad \textbf{(VI-17)}$$

From the values of $\Delta \overline{G}_i^\circ$ and ΔS° for a given reaction we can obtain ΔH° from the equation

$$\Delta \overline{G}_i^\circ = \Delta H^\circ - T\,\Delta S^\circ \qquad \textbf{(VI-18)}$$

The half reactions (VI-1) and (VI-2) discussed previously do not proceed individually, but rather they always proceed in pairs with other half reactions. From the point of view of bookkeeping it is convenient to assume the standard *emf* (E°) of one electrode to be zero and associate the *emf* **of the cell** with the other electrode. By common consent, the hydrogen electrode has been chosen as the standard electrode, and the standard *emf* of any other electrode is taken to be that measured in comparison with the hydrogen electrode. Accordingly, the E° measured for reaction (VI-3) (which is −0.222

or 0.222 volt, depending upon the sign convention used) is taken by convention to be that of the half reaction

$$Ag(s) + Cl^- \text{(soln.)} = AgCl + e^- \qquad E° = -0.222 \text{ volt} \qquad \textbf{(VI-19)}$$

while for the half reaction

$$H_2(g) = 2H^+ \text{(soln.)} + e^- \qquad E° = 0.00 \text{ volt} \qquad \textbf{(VI-20)}$$

by definition.

EXAMPLES AND APPLICATIONS

We shall now illustrate the use of thermodynamics and *emf* values in calculating the equilibrium concentrations of species in solutions at room temperature (298°K). The equilibria discussed are of several kinds, but in each case we shall ignore activity coefficients and assume that the activity is equal to the molality (practical description). In order to make this assumption a reasonable one we shall consider only solutions which are sufficiently dilute that the activity and molality are essentially equal, that is, sufficiently dilute that the solute obeys the appropriate form of Henry's Law.

Example 1. Consider the reaction in which solid Hg_2Cl_2 dissolves in water to form $Hg_2{}^{2+}$ and Cl^- ions in solution. We shall calculate the solubility of the Hg_2Cl_2. We write the balanced net reaction

$$Hg_2Cl_2(s) = Hg_2{}^{2+} \text{(soln.)} + 2\,Cl^- \text{(soln.)} \qquad \textbf{(VI-21)}$$

By referring to *Oxidation Potentials* (see Appendix I) we find for the mercury-mercurous ion couple

$$2\,Hg = Hg_2{}^{2+} + 2e^- \qquad E° = -0.789 \text{ volt} \qquad \textbf{(VI-22)}$$

and for the mercury-mercurous halide couple

$$2\,Cl^- + 2\,Hg = Hg_2Cl_2 + 2e^- \qquad E° = -0.268 \text{ volt} \qquad \textbf{(VI-23)}$$

Subtracting (VI-23) from (VI-22) yields

$$Hg_2Cl_2 = Hg_2{}^{2+} + 2\,Cl^- \qquad E° = -0.521 \text{ volt} \qquad \textbf{(VI-24)}$$

We thus obtain

$$\Delta\overline{G}_i° = (2)(23{,}060)(0.521) \text{ cal} \qquad \textbf{(VI-25)}$$

from which, using $\Delta\overline{G}_i° = -RT \ln K$, we obtain

$$K = m_{Cl^-}{}^2 m_{Hg_2{}^{2+}} = 1.1 \times 10^{-18} \qquad \textbf{(VI-26)}$$

By the stoichiometry of the solution reaction

$$2m_{Hg_2{}^{2+}} = m_{Cl^-} \qquad \text{(VI-27)}$$

and thus

$$4(m_{Hg_2{}^{2+}})^3 = 1.1 \times 10^{-18} \qquad \text{(VI-28)}$$

Solving for $m_{Hg_2{}^{2+}}$ we obtain 6.5×10^{-7} molal, a dilute solution indeed! We thus conclude that Hg_2Cl_2 is soluble to the extent of producing 6.5×10^{-7} mole of ${Hg_2}^{2+}$ per 10^3 grams of water and 13.0×10^{-7} mole of Cl^- per 10^3 gram of water at room temperature. The equilibrium constant for a solution reaction to produce a sparingly soluble salt is called a "solubility product" for the obvious reason that it is a product of activities (or in this case, as with many others, molalities) raised to the appropriate powers which provides a measure of the solubility of the salt.

Example 2. As a second example we shall calculate the extent of dissociation of a "weak electrolyte." A weak electrolyte is a substance which, in solution, exists in both the molecular form and the ionized form, with the former being greatly predominant. An example is H_2O. Water dissociates according to the reaction

$$H_2O(l) = H^+ \text{ (soln.)} + OH^- \text{ (soln.)} \qquad \text{(VI-29)}$$

We start by assuming that water is only very slightly dissociated (it is not a good electrical conductor when pure) and thus that it is meaningful to take the activity of molecular water as unity in pure water. We then examine the tables in *Oxidation Potentials* and obtain

$$H_2 = 2\,H^+ + 2e^- \qquad E^\circ = 0.00 \text{ volt} \qquad \text{(VI-30)}$$

by definition. We also find

$$H_2 + 2\,OH^- = 2\,H_2O + 2e^- \qquad E^\circ = 0.828 \text{ volt} \qquad \text{(VI-31)}$$

By subtracting reaction (VI-31) from (VI-30) we obtain

$$2\,H_2O(l) = 2\,H^+ \text{ (soln.)} + 2\,OH^- \text{ (soln.)} \qquad \text{(VI-32)}$$

for which

$$\begin{aligned} \Delta\overline{G}_i{}^\circ &= 0.828 \cdot 2 \cdot 23.060 \text{ kcal} \\ &= 38.19 \text{ kcal} \end{aligned} \qquad \text{(VI-33)}$$

The Gibbs free energy change for one mole of H_2O dissociating into hydrogen ions and hydroxide ions (solvated, of course) is one-half this value, or 19.05 kcal. It follows that

$$K = m_H m_{OH^-} = 10^{-14} \qquad \text{(VI-34)}$$

Thus water is, indeed, only slightly ionized at room temperature. We define pH as the negative logarithm of the activity of hydrogen ion. It follows that the pH of neutral water is seven.

Example 3. As a final example we consider the extent of an oxidation-reduction reaction between ions in solution. We have chosen the reaction between thiosulfate ion ($S_2O_3^{2-}$) and triiodide ion (I_3^-) —a chemical reaction which is important in analytical chemistry. The reaction is

$$2\,S_2O_3^{2-}\text{ (soln.)} + I_3^-\text{ (soln.)} = S_4O_6^{2-}\text{ (soln.)} + 3\,I^-\text{ (soln.)} \qquad \textbf{(VI-35)}$$

in which tetrathionate and iodide ions are the reaction products. Again from *Oxidation Potentials* we obtain

$$2\,S_2O_3^{2-} = S_4O_6^{2-} + 2e^- \qquad E^\circ = -0.08 \text{ volt} \qquad \textbf{(VI-36)}$$

and

$$3\,I^- = I_3^- + 2e^- \qquad E^\circ = -0.536 \text{ volt} \qquad \textbf{(VI-37)}$$

By subtracting (VI-37) from (VI-36) we obtain the balanced net reaction given above and

$$\begin{aligned} \Delta\overline{G}_i^\circ &= -2 \cdot 0.456 \cdot 23.06 \text{ kcal} \\ &= -21.0 \text{ kcal} \end{aligned} \qquad \textbf{(VI-38)}$$

We then obtain for the equilibrium constant

$$K = \frac{m_{I^-}^{3}\, m_{S_4O_6{}^{2-}}}{m_{I_3^-}\, m_{S_2O_3{}^{2-}}^{2}} = 3.2 \times 10^{15} \qquad \textbf{(VI-39)}$$

We note that the numerator of the equilibrium expression is very much larger than the denominator, and thus that the products are strongly favored over the reactants. This is one important criterion for determining whether or not a reaction is suitable for use in an analytical determination. Some other criteria are the absence of side reactions and the shelf life of the reagents. The first of these questions can again be examined by the use of thermodynamics, although the latter question is usually a kinetic, rather than a thermodynamic one.

SUMMARY

In this chapter we have considered the application of thermodynamics to reversible electrochemical cells. We have obtained the formulae by means of which important thermodynamic values can be obtained from *emf* measurements. We have discussed the solubility of a strong electrolyte, the dissociation of a weak electrolyte, and an oxidation-reduction reaction as examples of chemical reactions in solution for which *emf* values might be useful for obtaining equilibrium information. We obtained the following important equations:

$$\Delta \overline{G}_i = -nFE$$

$$E = E^\circ - \frac{RT}{nF} \ln \frac{a_L^{\,l} a_M^{\,m} \cdots}{a_A^{\,a} a_B^{\,b} \cdots}$$

$$\left(\frac{\partial E^\circ}{\partial T}\right)_P = \frac{\Delta S^\circ}{nF}$$

Problems

VI-1 Given (at 298°K)

$$Mg = Mg^{2+} + 2e^- \qquad E^\circ = 2.37 \text{ volts}$$

and

$$2\,OH^- + Mg = Mg(OH)_2 + 2e^- \qquad E^\circ = 2.69 \text{ volts}$$

calculate the solubility product of $Mg(OH)_2$. Express the concentration of Mg^{2+} in a solution in equilibrium with $Mg(OH)_2$ as a function of pH.

VI-2 Given (at 298°K)

$$Mn^{2+} + 4\,H_2O = MnO_4^- + 8\,H^+ + 5e^- \qquad E^\circ = -1.51 \text{ volts}$$

and

$$2\,H_2O = H_2O_2 + 2\,H^+ + 2e^- \qquad E^\circ = -1.77 \text{ volts}$$

determine the equilibrium constant for the reaction between H_2O_2 and Mn^{2+} to form MnO_4^- and H_2O.

VI-3 Given (at 298°K)

$$Ni + 6\,NH_3(aq) = Ni(NH_3)_6^{2+} + 2e^- \qquad E^\circ = 0.47 \text{ volt}$$

and

$$Ni = Ni^{2+} + 2e^- \qquad E^\circ = 0.250 \text{ volt}$$

calculate the ratio of moles of Ni^{2+} to moles of $Ni(NH_3)_6^{2+}$ in a solution 1 molar in NH_3 (assume ideal behavior for all species).

VI-4 Given (at 298°K)

$$Pt + H_2S(aq) = PtS + 2\,H^+ + 2e^- \qquad E^\circ = 0.30 \text{ volt}$$

and

$$H_2S(aq) = S + 2\,H^+ + 2e^- \qquad E^\circ = -1.41 \text{ volts}$$

Is PtS stable with respect to dissociation into the solid elements?

VI-5 Given (at 298°K)

$$Zn = Zn^{2+} + 2e^- \qquad E° = 0.763 \text{ volt}$$

and

$$Ag = Ag^+ + e^- \qquad E° = 0.799 \text{ volt}$$

Describe the final state of a solution which is initially 1 molal in Ag^+ and into which is added an excess of solid zinc.

VI-6 Given (at 298°K)

$$H_2S = S + 2\,H^+ + 2e^- \qquad E° = -0.141 \text{ volt}$$

and

$$S^{2-} = S + 2e^- \qquad E° = 0.48 \text{ volt}$$

calculate the degree of dissociation of H_2S in a solution which is formed by dissolving 0.01 mole of $H_2S(g)$ per 1000 grams of water.

VI-7 Given (at 298°K)

$$2\,H_2O + C_6H_5CH_3(l) = C_6H_5COOH(aq) + 6\,H^+ + 6e^-$$
$$E° = -0.21 \text{ volt}$$

and using the value for the $Mn^{2+} - MnO_4^-$ couple given in Problem VI-2, discuss the feasibility of using permanganate ion to oxidize toluene ($C_6H_5CH_3$) to form benzoic acid (C_6H_5COOH) in solution.

VI-8 For the reaction

$$NH_3(g) = NH_3(aq)$$

the standard Gibbs free energy change at 298°K is −2.39 kcal. Calculate the *emf* of the cell for which the cell reaction is

$$Ni + 6\,NH_3(aq) + 2\,H^+ = H_2(g) + Ni(NH_3)_6{}^{2+}$$

if the partial pressure of $NH_3(g)$ over the solution is 1/2 atm, the pH of the solution is 10 and the concentration of $Ni(NH_3)_6{}^{2+}$ is 1 molar. Use the E° data from Problem VI-3.

CHAPTER VII

Phase Diagrams

COMPONENTS

In this chapter we shall deal with the topic of heterogeneous equilibrium in systems with variable composition. While it is possible to deal with this topic in a mathematically analytic fashion [see Franzen and Gerstein, *A. I. Ch. E. Journal* **12** 364 (1966)], we shall present here only a discussion of diagrams which represent pictorially the equilibria in such systems. Such diagrams are called **phase diagrams.** Phase diagrams can be characterized according to the *minimum component number*, *c*, generally simply called the number of components.

The minimum component number can be defined as the minimum number of chemical substances from which each phase in the system can be *independently* synthesized. By *independently synthesized,* we mean that for each phase which is not a component we can write a chemical reaction involving only the phase and components. Consider the system involving the solid, liquid and vapor phases of water. This is clearly a one component system since we can take $H_2O(l)$ as the component and "synthesize" the phases according to

$$H_2O(l) \rightarrow H_2O(s) \tag{VII-1}$$

$$H_2O(l) \rightarrow H_2O(g) \tag{VII-2}$$

Next consider the system in which $CaCO_3(s)$, $CaO(s)$ and $CO_2(g)$ are in equilibrium. This is a two component system. Take, for example, $CaO(s)$ and $CO_2(g)$ to be the components. We can "synthesize" $CaCO_3(s)$ according to

$$CaO(s) + CO_2(g) = CaCO_3(s) \tag{VII-3}$$

which involves the formation of the solid $CaCO_3$ phase from the two components. It is clear that we can "synthesize" the other phases simply by taking the required number of moles of each component. Note that while it is possible to synthesize both $CaO(s)$ and $CO_2(g)$ from $CaCO_3(s)$ according to the reverse of reaction (3), the system is not a one component system because it is not possible to synthesize *independent* quantities of these two phases with a single component. By means of the reverse of reaction (VII-3) the number of moles of $CaO(s)$ formed is equal to the number of moles of $CO_2(g)$ formed.

One last point should be mentioned with regard to components of a system. This is that the number of components depends upon the chemical reactions that *can* occur. Consider, for example, the system made up of a mixture of $H_2O(g)$, $H_2(g)$ and $O_2(g)$. If a catalyst is present, or if the temperature is sufficiently high, then the reaction

$$H_2O(g) = H_2(g) + \tfrac{1}{2}\,O_2(g) \qquad \textbf{(VII-4)}$$

will come to equilibrium. In this case the system in question is a two component system (i.e., has a minimum component number of two) as discussed above. At room temperature, and in the absence of a catalyst, this reaction does not occur for kinetic reasons, and thus the system in question is a three component system. In the latter case the only method of obtaining an arbitrary mixture of $H_2O(g)$, $H_2(g)$ and $O_2(g)$ is by mixing the required amounts of the three components, whereas in the former case any of the three gaseous species could be synthesized by means of a chemical reaction involving only itself and the other two species. Thus the system in which reaction (VII-4) is kinetically feasible is a two component system, that in which it is not is a three component system. In general, only those reactions which are kinetically feasible should be considered for the synthesis of a phase in the system.

EUTECTIC MELTING

In Appendix IV we show that the equilibrium freezing point of a solvent is depressed by a solute which obeys Henry's Law according to the relation

$$\Delta T_f \cong \frac{N_2 R T_f^2}{\Delta \overline{H}_{fus}} \cong \frac{M_1 R T_f^2 M_2}{1000\,\Delta \overline{H}_{fus}} = K_f M_2 \qquad \textbf{(VII-5)}$$

provided the solute is insoluble in the solid.

If we plot N_1 versus T, first for the depression of the freezing point of A by solute B, and then for the depression of the freezing point of B by solute A, we obtain the diagram shown in Fig. 7.1a, where we

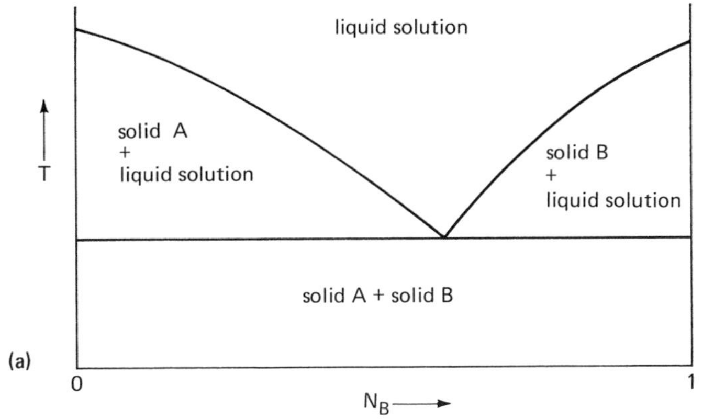

Figure 7.1 (**a**) Simple eutectic melting.

have terminated the curves at their intersection and drawn a horizontal line through the point of intersection. The diagram we obtain is called a **two-component phase diagram.** This diagram provides us with a picture of the behavior of the A – B system as a function of temperature and composition. A real system which very closely approximates the behavior of this system is the Bi – Cd system. The interpretation of the lines and areas of Fig. 7.1 is analogous to that of the one component phase diagram discussed in Chapter III, page 52, but requires a bit of further explanation. In the case of a two component system, we have instead of *P* and T as the only variables, the added variable of composition, or in this case, mole fraction, N_B. Therefore, while the one component diagram could be represented in two dimensions with *P* and T as axes, a full representation of a two component diagram would require a three dimensional figure, with axes *P*, T and N_B. One way of avoiding this difficulty is to make constant pressure or constant temperature "cuts" of such a three dimensional figure, and limit the variables to either *P* and N_B or T and N_B. The latter representation is shown in Fig. 7.1a. For a condensed system such as considered in Fig. 7.1a, the pressure dependence at ordinary pressures is very small, i.e., the curves from one to one hundred atmospheres usually look very similar, so not much information is lost by considering the constant pressure cut at one atmosphere.

As with the one component phase diagram, each point on a line or in a region represents a definite state of the system. However, an area represented a one phase region in a one component diagram, whereas areas can represent either two phase regions *or* one phase regions in a two component diagram, and lines can represent three phase regions. For example, the area labeled "liquid solution" in Fig. 7.1a is a one phase region. This means that at each point, specified by a temperature, pressure (fixed throughout the entire region) and composition in this area, the system A – B will consist of a homogeneous liquid solution. The area labeled "solid A + liquid solution," however, is a *two phase* region, the two phases coexisting in equilibrium being pure solid A and liquid of some composition. The composition in question is specified by the temperature of the system. At a given temperature, say T'', pure A at temperature T'' will coexist in equilibrium with the solution at T'', and the composition of the solution under such conditions is clearly, N'_B. This is to say that if the state of the system is such that its temperature and composition constitute a point within the region labeled "solid A + liquid solution," then the "system" will physically be a solid and a liquid, the solid being pure A, and the liquid being a solution of A and B, the composition of the solution being specified by the temperature at which the liquid and solid co-exist in equilibrium.

A distinction should clearly be made between the *overall* composition of the system, and the compositions of the two phases which co-exist in equilibrium. For example, at temperature T'' in Fig. 7.1b and any overall composition between $N_B = 0$ and $N_B = N'_B$ the two phases co-existing in equilibrium will *always* be pure A and liquid solution with mole fraction N'_B. A line such as the horizontal line at T'' in Fig. 7.1b which specifies constant temperature and connects two one phase regions (in this case pure A and liquid solution) is called a **tie line.** A tie line is a simple construction which "ties" together the two phases which co-exist in equilibrium. The region labeled "solid B + liquid solution" in Fig. 7.1a is a two phase region any point of which describes a state of the system in which pure solid B co-exists in equilibrium with a liquid solution of A and B, the composition of the solution being fixed by the temperature. Similarly, the area labeled "solid A + solid B" is a two phase region consisting of solid A and solid B.

The horizontal line between the region labeled "solid A + solid B"

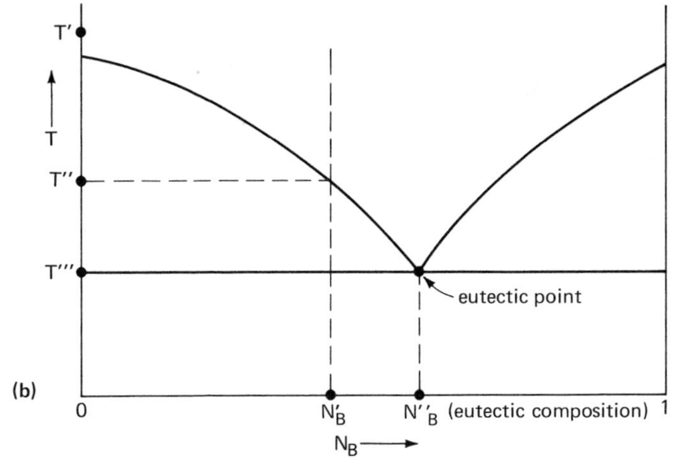

Figure 7.1 (**b**) Simple eutectic melting with construction to illustrate phases which co-exist in equilibrium and eutectic composition.

and the regions immediately above it is particularly interesting, as it represents a three-phase region. The system at the temperature T‴ (Fig. 7.1b) and with any overall composition between $N_B = 0$ and $N_B = 1$ will consist of three phases co-existing in equilibrium. It is immediately obvious which three phases these are from an inspection of Fig. 7.1b. They are solid A, solid B, and liquid solution with composition N'''_B. The intersection of the two "liquidus" lines at T‴ defines the composition N''_B, and this composition is called the **eutectic composition.** The point specified by the temperature T‴ and the composition N''_B is called the **eutectic point.** A phase diagram such as shown in Fig. 7.1b is called a simple eutectic diagram.

We can illustrate the utility of phase diagrams such as Fig. 7.1b in discussing the behavior of the system by considering the changes which occur as the temperature of a system with the composition N'_B is cooled from T′. We call your attention to the vertical dashed line on Fig. 7.1b. We note that as the system is cooled from T′ to T″ the only change is in the temperature of the liquid solution. At T″, however, the first precipitation of pure solid A occurs. Between T″ and T‴, A continues to precipitate from the solution according to

$$\text{A (in solution)} = \text{A}(s) \qquad \textbf{(VII-6)}$$

i.e., at all temperatures between T‴ and T″ there are two phases in equilibrium–namely, solid A and liquid solution, as has been discussed above. In this temperature region the composition of the liquid varies according to equation (A-IV-13), the equation of the curved line which is the upper boundary of the two-phase region.

At T‴ the three phases A(*s*), B(*s*), and solution coexist in equilibrium. The composition of the solution when this equilibrium obtains is, as described above, that of the intersection of the two freezing-point depression curves and is the eutectic composition. There is a chemical reaction which occurs involving the three phases co-existing in equilibrium at T‴. We can decide what this chemical reaction is in the following way. First, we look at the two phase region on the dashed vertical line immediately above the T‴ (eutectic temperature) line and note that the two phases coexisting in equilibrium prior to the three phase reaction are liquid and A(*s*). We next similarly examine the two phase region below the eutectic line and note that the two phases coexisting in equilibrium after the three phase reaction are A(*s*) and B(*s*). It is clear then that liquid is consumed in the reaction, and B(*s*) is produced. There are, apparently, two possible chemical reactions which occur along the eutectic line, namely

$$\text{liquid} + A(s) = B(s) \qquad \textbf{(VII-7)}$$

and

$$\text{liquid} = A(s) + B(s) \qquad \textbf{(VII-8)}$$

But equation (VII-7) can not be balanced since both liquid and A(*s*) are richer in A than is B(*s*) (which in this case contains no A at all). Thus the reaction, called the eutectic reaction, which occurs at the eutectic temperature is that of equation (VII-8). Of course we could replace the word "liquid" with a formula $AB_y(l)$ which represents the eutectic composition and balance the equation. This would yield

$$AB_y(l) = A(s) + yB(l) \qquad \textbf{(VII-9)}$$

SOLID SOLUBILITY

It sometimes occurs that solids are soluble in each other, just as liquids are. Solid solutions are known for many pairs of substances. For example, it is known that Si and Ge are completely soluble in each other, both in the solid and liquid states. The temperature-composition phase diagram for such a pair of substances, if both the liquid and the solid solutions obey Henry's Law, has the appearance shown in Fig. 7.2. We see from this diagram, by an analysis similar

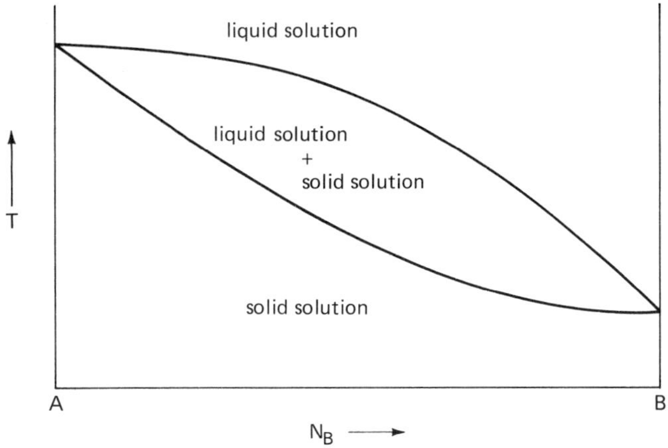

Figure 7.2 Solid-liquid phase diagram in which both components obey Henry's Law; complete solid solution and liquid solution.

to that given in the previous case, that when a given liquid solution is cooled a temperature is reached at which solid solution precipitates, and the solid solution continues to precipitate over a temperature range given by the corresponding vertical width of the two phase region. Both the liquid and solid solution concentrations vary during the solidification process in such a system. The concentrations of the two phases are given at any temperature by the vertical tie line corresponding to that temperature.

Not all solid solutions remain as solutions at low temperatures. It often happens that the solid solution separates into two immiscible portions when the temperature is lowered. Such behavior is indicated by the phase diagram shown in Fig. 7.3. To determine the behavior of a given solid solution when it is cooled one again considers a vertical line corresponding to the composition of the solid solution. If the vertical line does not intersect the line enclosing the two phase region then it is known that the given solid solution remains miscible at the temperatures of interest. If, on the other hand, the vertical line does intersect the line enclosing the two phase region, then it is known that the solid solution separates into two solid phases at equilibrium. The two phases which co-exist in equilibrium at a given temperature can be determined by the use of the tie line appropriate to that temperature, in a manner analogous to those discussed in the cases of other tie lines in two phase regions.

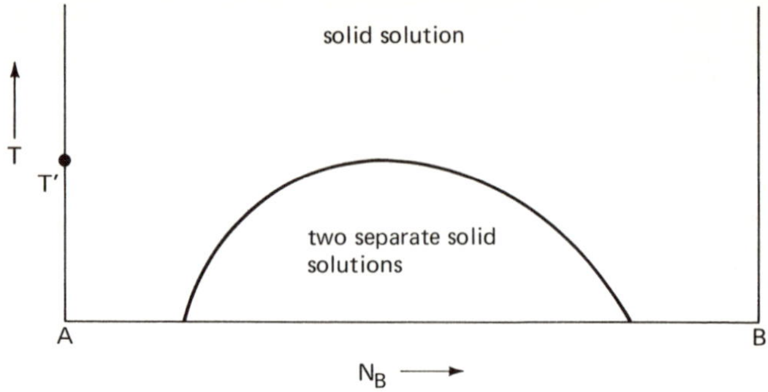

Figure 7.3 System exhibits complete solid solution above temperature T′, but separates into two solid solutions of different composition below T′.

It is observed that many solutions do not obey Henry's Law at finite concentrations. In this case, the phase boundary lines such as shown in Fig. 7.2 become distorted, and may have the form shown in Fig. 7.4. This figure is the phase diagram of a solid solution which melts congruently. Congruent melting occurs when the liquid and the solid have the same composition. If both the behavior illustrated by Fig. 7.3 (limited solid-solid miscibility) and that illustrated by Fig. 7.4 (nonideal solution behavior at finite concentrations) occur, then the phase diagram shown in Fig. 7.5 results.

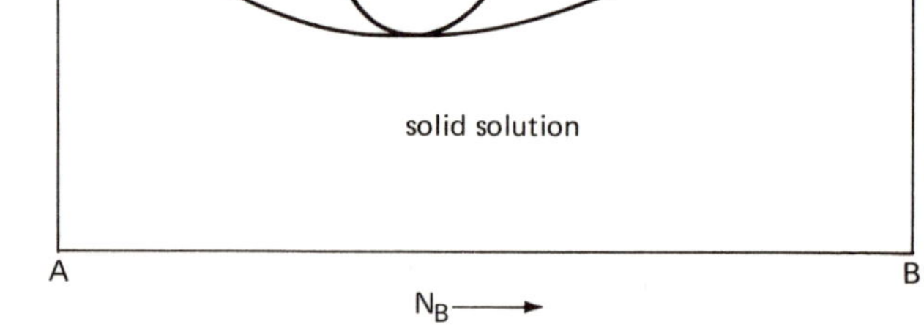

Figure 7.4 Liquid solution-solid solution equilibria in which components do not obey Henry's Law.

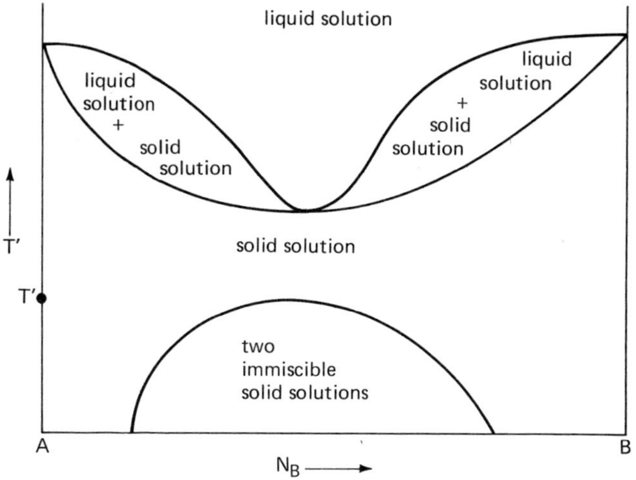

Figure 7.5 Neither component obeys Henry's Law in the melt, and solid solution-solid solution separation occurs below T′.

We can now consider what happens if the solid miscibility gap extends up to the point where it intersects the liquid-solid equilibrium lines. In this case we obtain the behavior diagrammed in Fig. 7.6, which is simply a eutectic diagram with solid solution (compare with Fig. 7.1), which is a very common type of phase diagram.

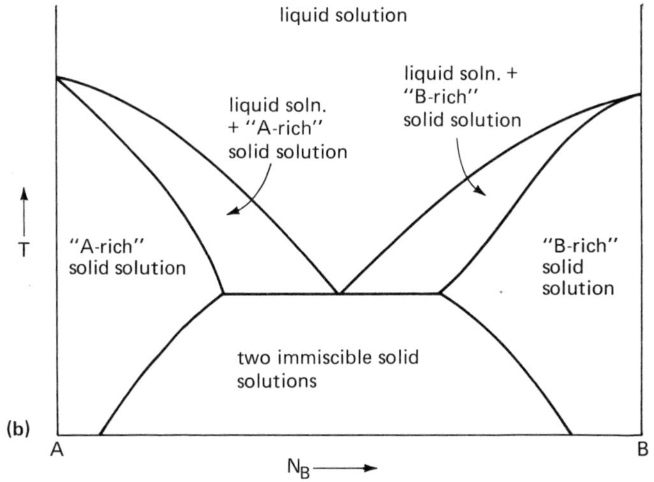

Figure 7.6 Solid solution-solid solution immiscibility gap meets solid solution-liquid solution region.

COMPOUND FORMATION

It sometimes occurs that compounds are formed between the components of a system. Such compounds are formed in binary oxide systems, which are extensively studied by ceramists, and binary metal systems, such as those studied by metallurgists. An example of a mixed oxide compound is mullite, Al_6SiO_{11}, which is formed from 3 moles of alumina (Al_2O_3) and 1 mole of silica (SiO_2). An example of a binary metal compound is Cu_9Al_4.

When compounds such as AB_y are formed from components A and B, it is possible to consider two separate phase diagrams, namely the AB_y-A diagram and the AB_y-B diagram. Each of the diagrams might, for example, be of the form of Fig. 7.6, as shown in Fig. 7.7. If we then consider the A-B phase diagram, it will simply consist of the combination of the diagrams of Figs. 7.7a and 7.7b, as shown in Fig. 7.8.

We call this diagram one of a system with a single, congruently melting compound. The melting phenomenon at the maximum of the T-N curve for AB_y-liquid equilibrium is analogous to that shown in Fig. 7.4, except that the congruent melting is at a maximum in the T-N curve of Fig. 7.8 rather than a minimum.

(a)

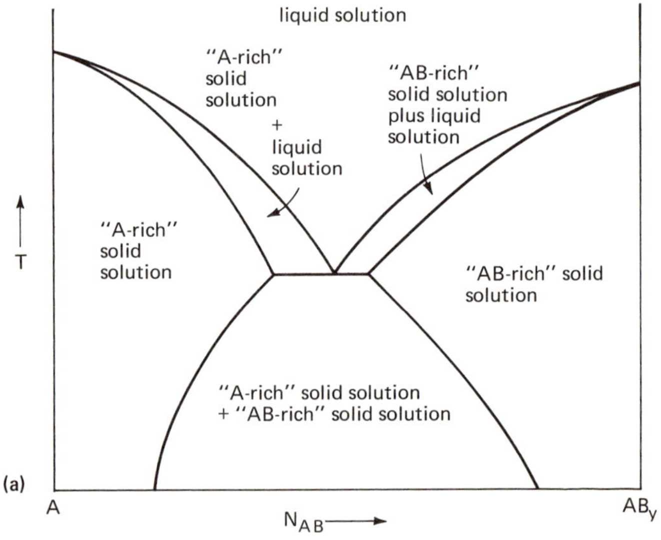

Figure 7.7 **(a)** Compound formation in which compound and pure component both have solid solubility.

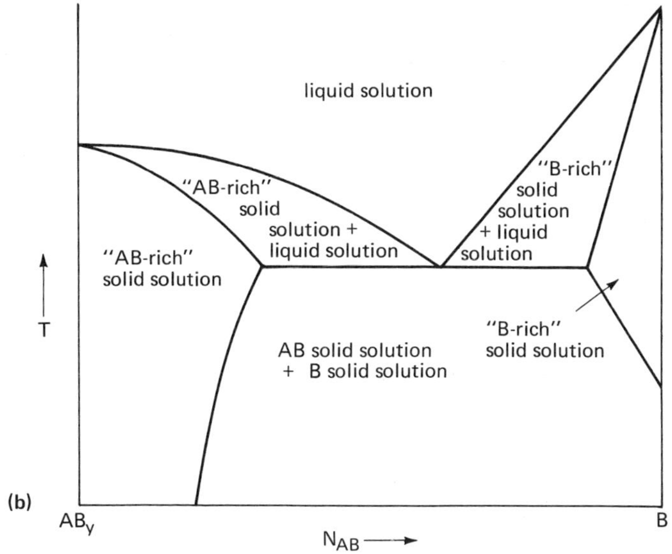

Figure 7.7 (**b**) Compound formation in which compound and pure component have solid solubility.

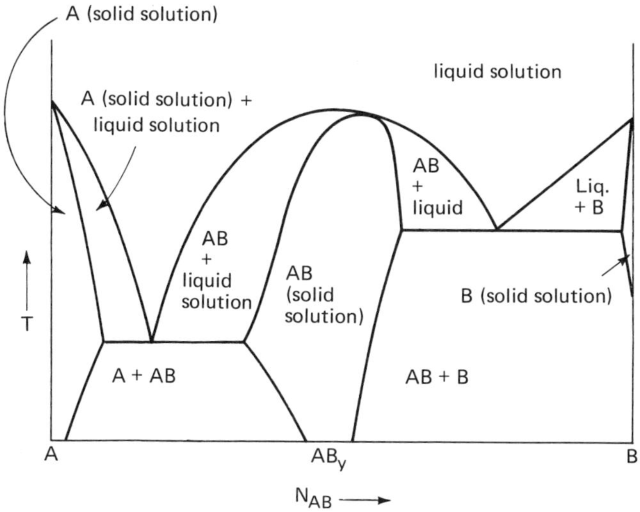

Figure 7.8 Combination of phase diagrams 7.7a and 7.7b illustrating choice of components as either pure elements (A and B) or of element (A or B) and compound (in this case AB with a large range of solid solution with both A and B).

INCONGRUENT MELTING OF COMPOUNDS

Since up to the present we have specifically denoted the melting of some compounds as "congruent" you have no doubt guessed that there are "incongruently" melting compounds. Such is indeed the case. Examples of incongruent melting are found when a compound AB_y melts to yield another compound, AB_x, plus melt, or one of the components, such as A(s), plus melt,

$$AB_y(s) = AB_x(s) + \text{melt}$$
$$AB_y(s) = A(s) + \text{melt}$$

Such melting phenomena result in phase diagrams as shown in Figs. 7.9a and 7.9b, where Fig. 7.9a represents incongruent melting of a compound $A_2B(s)$ with negligible solid solubility, and Fig. 7.9b represents incongruent melting of $AB_y(s)$ with observable solid solubility of the components in the compound.

It is possible to employ the technique previously used in this chapter to determine what chemical reaction occurs along a three

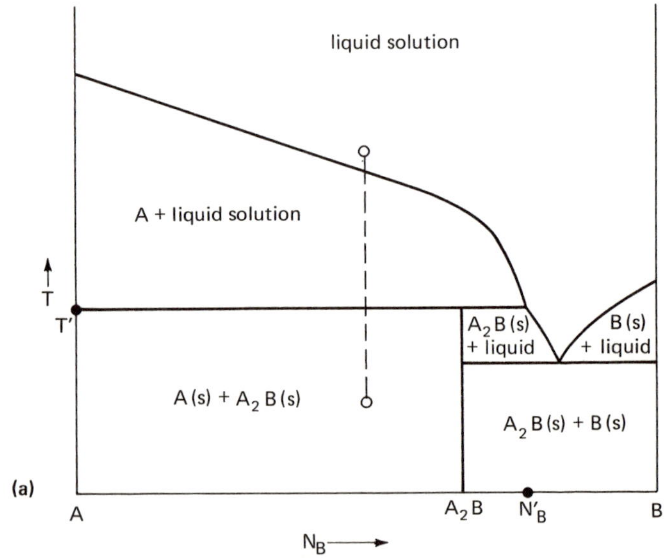

Figure 7.9 (a) Incongruent melting of a "line compound" (i.e., a compound without a range of composition) A_2B. At T′, $A_2B(s)$ decomposes according to the reaction $A_2B(s) = A(s) +$ liquid solution with composition N_B'.

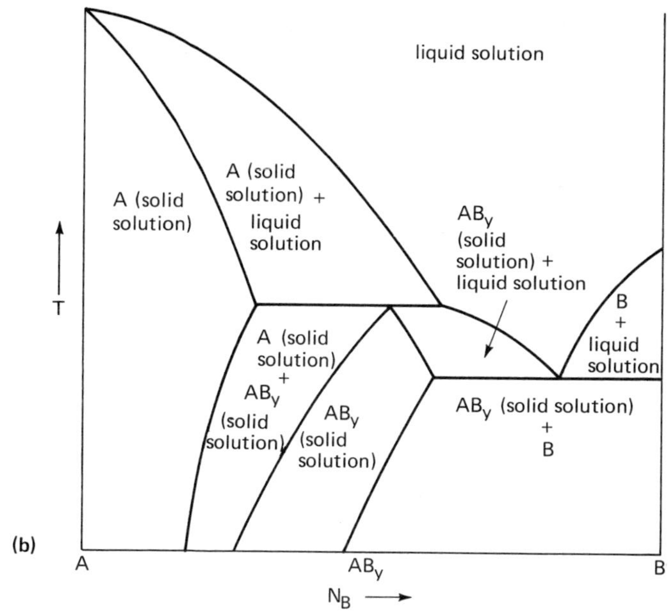

Figure 7.9 **(b)** Incongruent melting of a compound AB_y with a range of composition.

phase line corresponding to the melting (or precipitation) of the compound. For example, consider cooling the system along the vertical dashed line in Fig. 7.9a. At temperatures slightly higher than that of the three-phase equilibrium, which involves A(s), A_2B(s) and melt, A co-exists in equilibrium with liquid solution. After the reaction has occurred A_2B(s) and A(s) remain. Clearly, then, liquid solution has been consumed and A_2B(s) has been produced in the solidification reaction. The only way in which the equation involving A_2B(s), A(s) and liquid can be balanced such that liquid is consumed and A_2B(s) produced is

$$A(s) + \text{liquid} = A_2B(s) \tag{VII-10}$$

This is the reaction which occurs when the system is cooled through the incongruent melting temperature of the compound A_2B.

Clearly a system might have many compounds, some of which melt congruently, some of which melt incongruently, some of which exhibit significant solid solution and some of which do not. Thus it is possible to obtain complex phase diagrams, such as shown in Fig. 7.10, which are simply combinations of the cases discussed above.

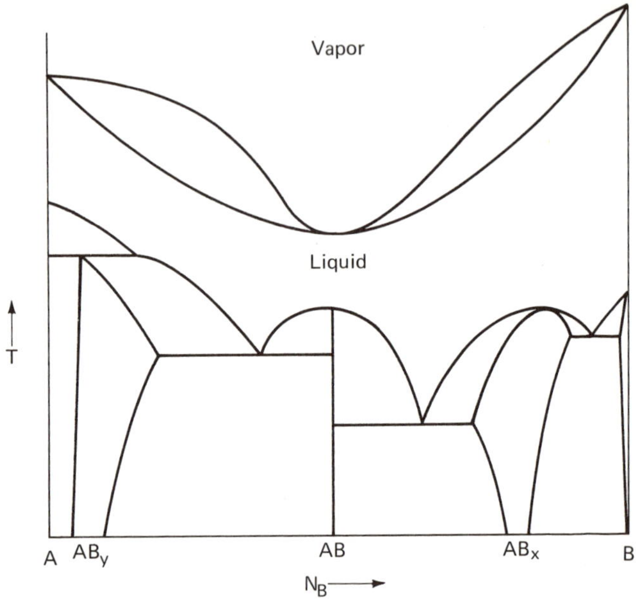

Figure 7.10 Some possible combinations of the cases discussed and illustrated in Figs. *7.1–7.9*.

THE PHASE RULE

In earlier chapters we have made use of the fact that for a single component closed system the intensive state of the system is fixed by two variables. If we wish to express the quantity of matter in the system as well we must fix one more variable. In fact for each additional component of a multicomponent system we must fix one more variable, or c + 2 variables in all. Now these variables are of two kinds—there are those such as T, *P*, and N_i, which fix the intensive state of the system and those which fix the mass of each phase, thereby fixing the extensive state of the system. Let us call the number of independent intensive variables f and the number of masses (i.e., the number of phases) p. We then have

$$f + p = c + 2 \tag{VII-11}$$

or

$$f = c - p + 2 \tag{VII-12}$$

That is, the number of independent intensive variables is equal to the number of components minus the number of phases plus two, which is the **Gibbs Phase Rule.**

This important relation enables us to determine the number of variables we must or may specify in a given experiment. If we were to examine the system discussed previously in this chapter consisting of $CaCO_3$(s), CaO(s) and CO_2(g), for example, we would find that the two components and three phases imply that there is one independent intensive variable. Thus the pressure is a function of the temperature of this system, and we need fix only the temperature to obtain a fixed pressure. On the other hand if we were to examine a one-component system, such as that formed by taking a given amount of H_2O, in which there are three phases co-existing then the number of independent intensive variables would be zero. That is the point at which solid, liquid and vapor are in equilibrium in a one component system is a fixed point determined only by the nature of the component.

Under conditions at which liquid and vapor coexist in a one component system the number of independent intensive variables is one. That is, the vapor pressure of a single component liquid is a function only of the temperature of the system. We have seen this previously, for the Clausius-Clapeyron equation (p. 54) relates the vapor pressure of a liquid to its temperature.

SUMMARY

In this chapter we have extended our discussion of phase diagrams to systems with variable composition. We have introduced the concept of minimum component number and developed the Gibbs Phase Rule. We have introduced the concepts of eutectic melting, solid solubility, compound formation, congruent melting and incongruent melting. We have demonstrated the use of tie and vertical construction lines in interpreting phase diagrams.

Problems

VII-1 For each system described below give c, the number of components and f, the number of independent intensive variables:

a. $Al_2O_3(s)$ in equilibrium with $Al_2O_3(l)$.

b. $Al_2O_3(s)$ in equilibrium with $Al(g)$, $O_2(g)$, $O(g)$, $Al_2O(g)$ and $AlO(g)$ such that the condensed phase and vapor phase have different compositions. What if the overall vapor composition is the same as the solid?
c. $Al_2O_3(s)$ in equilibrium with $Al(l)$.
d. $CO_2(s)$ in equilibrium with $CO_2(g)$ at low temperatures.
e. $CO_2(g)$ in equilibrium with $CO(g)$, $O(g)$, $O_2(g)$, $C(g)$, $C_2(g)$ and $C_3(g)$ at elevated temperatures.

VII-2 The temperature-composition phase diagram of the Mg-Pb system is given below. Note that the composition is given in weight percent.

a. Determine the stoichiometry of the compound.
b. What is the melting point of the compound?
c. What reaction occurs at the eutectic temperature when a 60% Pb sample is heated?
d. What is the solubility of Pb in a Mg-Pb solution at 500°C?

VII-3 The temperature-composition phase diagram for the Ge-Si system is given below. Describe what happens when an equimolar mixture of Ge and Si is cooled from 1400°C to 1000°C.

VII-4 The temperature-composition phase diagram for the Co-Sb system is given below.

a. Label the regions.
b. Describe what happens when a $N_{Sb} = 0.2$ melt is cooled from 1500°C to 1000°C.
c. Describe what happens when a $N_{Sb} = 0.7$ melt is cooled from 1300°C to 500°C.

VII-5 The temperature-composition phase diagram for the SiO_2-Al_2O_3 system is given below. Note that weight percent SiO_2 is plotted *vs.* temperature in degrees celsius.

a. What is the compound formed?
b. What occurs when this compound melts?
c. What is the solubility of Al_2O_3 in a fused Al_2O_3-SiO_2 melt at 2000°C?

VII-6 The temperature-composition phase diagram for the Mg-Ni system is given below.

a. What compounds exist in the Mg-Ni system?
b. Which melts congruently and which melts with decomposition?
c. What phases are present when equimolar amounts of Ni and Mg are heated to 600°C? to 1200°C? to 1800°C?

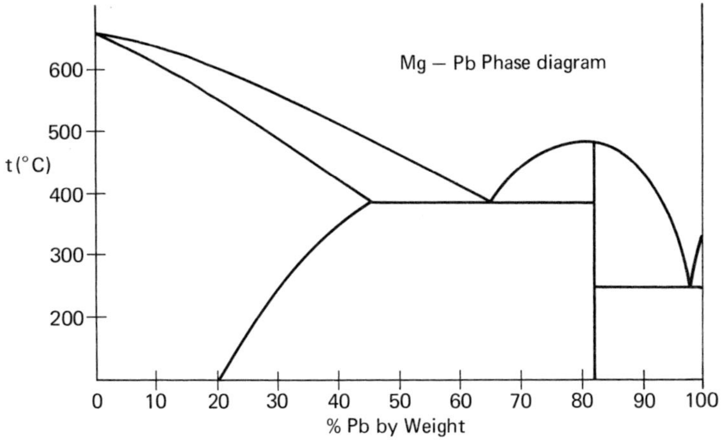

Figure for Problem VII–2.

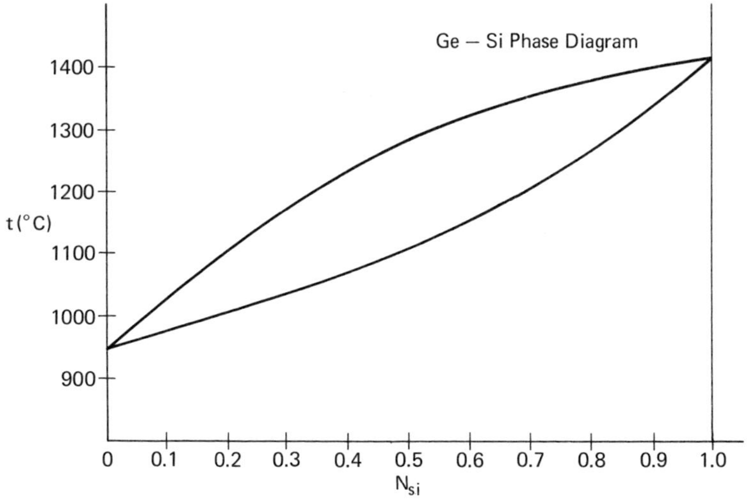

Figure for Problem VII-3.

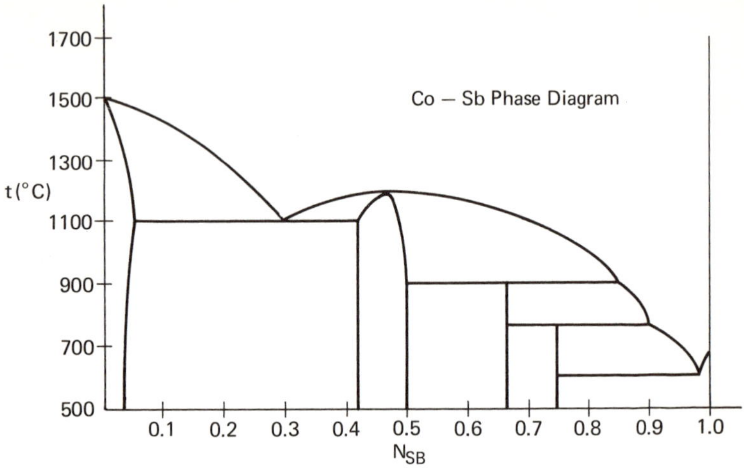

Figure for Problem VII-4.

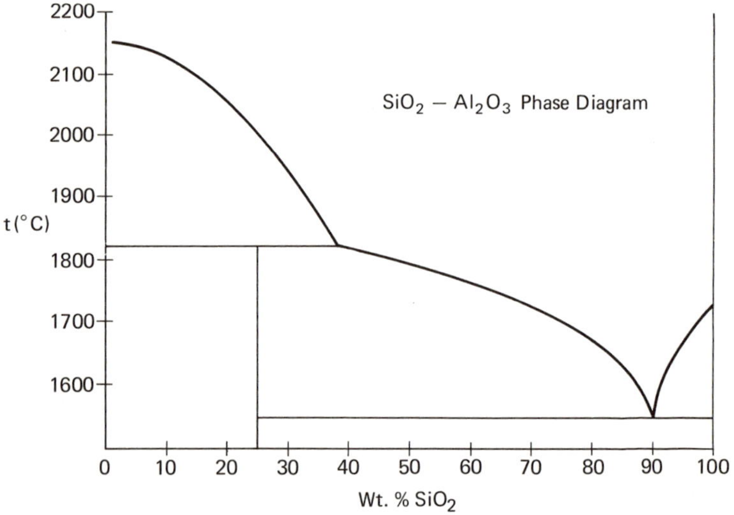

Figure for Problem VII-5.

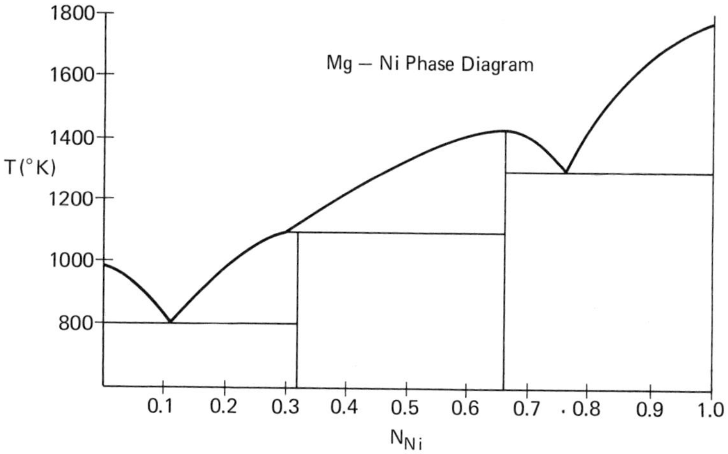

Figure for Problem VII-6.

Appendix I

Compilations of Thermodynamic Data

1. JANAF *Thermochemical Tables*

   Compiled at The Thermal Research Laboratory at the Dow Chemical Company under the direction of D. R. Stull. This compilation contains $C_p°$, $S°$, fef, $H° - H°_{298}$, $\Delta H_f°$, $\Delta G_f°$, and $\log K_p$ as functions of T in intervals of 100 Celsius degrees for many solid, liquid and gaseous elements and inorganic compounds.

2. *Thermodynamic Properties of the Elements*

   No. 18 of the Advances of Chemistry Series of the American Chemical Society, by D. R. Stull and G. C. Sinke (1956). This book contains $C_p°$, $H_T° - H°_{298}$, $S_T°$, fef, $\Delta H_f°$, $\Delta G_f°$, and $\log K_p$ for 92 elements in all states of aggregation.

3. *Selected Values of Thermodynamic Properties of Metals and Alloys*

   By Ralph Hultgren, Raymond L. Orr, Philip D. Anderson, and Kenneth K. Kelley, published by John Wiley and Sons, Inc., N.Y. (1963). This compilation consists of two parts: properties of pure metals and properties of binary alloys. The elemental properties listed include, $C_p°$, $H_T° - H°_{298}$, $S_T° - S°_{298}$, fef, vapor pressures and heats of transitions. The alloy properties listed include $\Delta G_f°$, $\Delta H_f°$ and activities. Phase diagrams are presented for the binary systems.

4. *Oxidation Potentials, Second Edition*

   By Wendell M. Latimer, published by Prentice-Hall, Inc., N.Y. (1952). This compilation contains $\Delta H_f°$, $\Delta G_f°$ and $S°$ at 298°K for many elements and inorganic compounds in various states of aggregation and, particularly, in solution.

5. Appendix 7 of *Thermodynamics, Second Edition*

By G. N. Lewis and M. Randall, revised by K. S. Pitzer and L. Brewer, McGraw-Hill Book Company, Inc., N.Y. (1961). This compilation contains fef's and $\Delta H°$ values for many inorganic and a few simple organic compounds.

6. *Contributions to the Data on Theoretical Metallurgy XIII*

Bureau of Mines Bulletin 584 by K. K. Kelley (1960) contains data on high temperature heat-content, heat capacity, and entropy for elements and inorganic compounds.

7. *Contributions to the Data on Theoretical Metallurgy XIV*

Bureau of Mines Bulletin 592 by K. K. Kelley (1961) contains data on low temperature heat capacities and entropies at 298°K of the elements and inorganic compounds.

8. *Selected Values of Physical and Thermodynamic Properties of Hydrocarbons and Related Compounds*

By F. D. Rossini, et al., Carnegie Press, Pittsburgh (1953), contains fef's for a large variety of hydrocarbons.

9. *National Bureau of Standards Circular 500*

By F. D. Rossini, D. Wagman, W. Evans, S. Levine and D. Jaffe (1952), contains enthalpies, entropies, and free energies of formation at 298°K for many elements and a wide variety of compounds in various states of aggregation.

APPENDIX II

Demonstration that the Internal Energy of an Ideal Gas is Independent of Volume at Constant Temperature.

What we must prove is that $\partial U/\partial V)_T = 0$. We start by defining a property A, called the **Helmholz Free Energy**, according to

$$A = U - TS \qquad \textbf{(A-II-1)}$$

Differentiation of this equation yields

$$dA = dU - \mathrm{T}\,dS - S\,d\mathrm{T} \tag{A-II-2}$$

and, since $dU = \mathrm{T}\,dS - P\,dV$ for a single component closed system, we have

$$dA = -S\,d\mathrm{T} - P\,dV \tag{A-II-3}$$

Now we can take A to be a function of T and V and thus

$$d\overline{A} = \left.\frac{\partial\overline{A}}{\partial\mathrm{T}}\right)_V d\mathrm{T} + \left.\frac{\partial\overline{A}}{\partial\overline{V}}\right)_T d\overline{V} \tag{A-II-4}$$

By equating the coefficients of the differentials in equation (A-II-3) and (A-II-4) we obtain

$$\left.\frac{\partial\overline{A}}{\partial\mathrm{T}}\right)_V = -S \quad \text{and} \quad \left.\frac{\partial\overline{A}}{\partial\overline{V}}\right)_T = -P \tag{A-II-5}$$

We can further differentiate equation (A-II-5) to obtain equation (A-II-6)

$$\frac{\partial^2\overline{A}}{\partial\mathrm{T}\partial V} = -\left.\frac{\partial\overline{S}}{\partial\overline{V}}\right)_\mathrm{T} \quad \text{and} \quad \frac{\partial^2\overline{A}}{\partial V\partial\mathrm{T}} = -\left.\frac{\partial P}{\partial\mathrm{T}}\right)_V \tag{A-II-6}$$

that is, the second partial derivative of A with respect to T at a constant V and then to V at constant T is equal to the negative of the partial of S with respect to V at constant T, and a similar statement can be made for the second equation. It is a result of calculus that

$$\frac{\partial^2\overline{A}}{\partial\mathrm{T}\partial V} = \frac{\partial^2\overline{A}}{\partial V\partial\mathrm{T}}$$

i.e., so long as the independent variables are the same the order of integration is immaterial. Thus we have demonstrated that

$$\left.\frac{\partial\overline{S}}{\partial\overline{V}}\right)_\mathrm{T} = \left.\frac{\partial P}{\partial\mathrm{T}}\right)_{\overline{V}} \tag{A-II-7}$$

We now examine the differential expression for U

$$d\overline{U} = \mathrm{T}\,d\overline{S} - P\,d\overline{V} \tag{A-II-8}$$

and note that this implies that

$$\left.\frac{\partial\overline{U}}{\partial\overline{V}}\right)_\mathrm{T} = \mathrm{T}\left.\frac{\partial\overline{S}}{\partial\overline{V}}\right)_\mathrm{T} - P \tag{A-II-9}$$

We substitute, according to (A-II-7)

$$\left.\frac{\partial\overline{U}}{\partial\overline{V}}\right)_\mathrm{T} = \mathrm{T}\left.\frac{\partial P}{\partial\mathrm{T}}\right)_V - P \tag{A-II-10}$$

and obtain an equation the right hand side of which is in terms of P, V and T and which can therefore be evaluated if the equation of state is known. Since we are discussing an ideal gas we have

$$P\bar{V} = \mathrm{RT}$$

and thus

$$\left.\frac{\partial P}{\partial \mathrm{T}}\right)_{\bar{V}} = \frac{\mathrm{R}}{\bar{V}} \qquad \text{(A-II-11)}$$

Substitution of this equation into (A-II-10) provides our desired result

$$\left.\frac{\partial \bar{U}}{\partial \bar{V}}\right)_{\mathrm{T}} = \frac{\mathrm{RT}}{\bar{V}} - P = 0 \qquad \text{(A-II-12)}$$

Appendix III

The Gibbs Free Energy and Chemical Potential of Ideal Gases in Mixtures

One characteristic of an ideal gas is that the internal energy and enthalpy of the gas do not depend upon the volume (or pressure) independently, but only upon the PV product, i.e., upon T (see App. II). A nonthermodynamic interpretation of this fact results from kinetic theory. It is that a gas which obeys $P\bar{V} = \mathrm{RT}$ is one for which the volume occupied by the gas itself is negligibly small (point particles) and for which the only interactions are elastic collisions. A mixture of such gases will behave in a similar fashion, and thus the enthalpy of a mixture of ideal gases is the same as that of the sum of the enthalpies of the pure gases, i.e., the enthalpy of mixing of ideal gases is zero.

Let us consider the Gibbs free energies, G_{A} and G_{B}, of two separated gases, A and B. When the gases are mixed, the Gibbs free energy of the mixture is

$$G = G_{\mathrm{A}} + G_{\mathrm{B}} + \Delta G_{\mathrm{mix}} \qquad \text{(A-III-1)}$$

But, since $\Delta H_{\mathrm{mix}} = 0$

$$\Delta G_{\mathrm{mix}} = -\mathrm{T}\,\Delta S_{\mathrm{mix}} \qquad \text{(A-III-2)}$$

Thus

$$G = G_{\mathrm{A}} + G_{\mathrm{B}} - \mathrm{T}\,\Delta S_{\mathrm{mix}} \qquad \text{(A-III-3)}$$

If we take the unmixed gases to be in their standard states,

$$G = n_A\overline{G}_A^\circ + n_B\overline{G}_B^\circ - T\Delta S_{mix} \tag{A-III-4}$$

But ΔS_{mix} is exactly the entropy required to change both gases from their standard states to their partial pressures in the mixture since no heat effects accompany the reversible mixing of the gases. This entropy change is given for one mole of each gas by equation (III-30), and thus the Gibbs free energy of the mixture is given by

$$G = n_A\overline{G}_A^\circ + n_B\overline{G}_B^\circ + n_A RT \ln P_A + n_B RT \ln P_B \tag{A-III-5}$$

The Gibbs free energy per mole of component A in the mixture and the chemical potential of component A in the mixture are both given by

$$\left.\frac{\partial G}{\partial n_A}\right)_{T_1 P, N_B} = \overline{G}_A \tag{A-III-6}$$

By differentiation of equation (A-III-5) according to equation (A-III-6)

$$\overline{G}_A = \overline{G}_A^\circ + RT \ln P_A \tag{A-III-7}$$

which is the same expression as that for the chemical potential and the Gibbs free energy per mole of pure ideal gas A at T and P_A. A similar expression

$$\overline{G}_B = \overline{G}_B^\circ + RT \ln P_B \tag{A-III-8}$$

holds for gaseous B.

Appendix **IV**

Temperature Dependence of Gibbs Free Energy and Derivation of Boiling Point Elevation and Freezing Point Elevation Relationships

DEPRESSION FORMULAE USING CHEMICAL POTENTIALS

In order to derive rigorously the freezing-point lowering and boiling-points elevation relations, it is convenient to use chemical potentials. It will be necessary to investigate the temperature dependence of ΔG in order to do so.

Temperature Dependence of ΔG

$$G = H - TS \qquad \textbf{(A-IV-1)}$$

By the Gibbsian equation $dG = +V\,dP - S\,dT$, we have

$$\left.\frac{\partial G}{\partial T}\right)_P = -S \qquad \textbf{(A-IV-2)}$$

(A-IV-1) and (A-IV-2) yield

$$G = H + T\left.\frac{\partial G}{\partial T}\right)_P$$

or

$$\Delta G = \Delta H + T\left.\frac{\partial\,\Delta G}{\partial T}\right)_P \qquad \textbf{(A-IV-3)}$$

Thus

$$\Delta H = \Delta G - T\left.\frac{\partial\,\Delta G}{\partial T}\right)_P$$

which is equivalent to

$$-\frac{\Delta H}{T^2} = \partial\left.\frac{(\Delta G/T)}{\partial T}\right)_P \qquad \textbf{(A-IV-4)}$$

Equation A-IV-4 is known as the Gibbs-Helmholtz equation.

Boiling Point Elevation

At equilibrium between a solution containing a nonvolatile solute and a solvent and the vapor of pure solvent

$$\overline{G}_1{}^{\circ,V} = \overline{G}_1{}^{L} \qquad \textbf{(A-IV-5)}$$

where the superscripts V and L refer to vapor and liquid, respectively, and the subscript 1 refers to solvent. The superscript $^\circ$ refers, as usual, to pure component. In terms of the activity of the solvent, we have

$$\overline{G}_1{}^{\circ,V} = \overline{G}_1{}^{\circ,L} + RT\ln a_1 \qquad \textbf{(A-IV-6)}$$

or

$$\frac{\overline{G}_1{}^{\circ,V} - \overline{G}_1{}^{\circ,L}}{T} = \frac{\Delta G_{\text{vap}}}{T} = R\ln a_1 \qquad \textbf{(A-IV-7)}$$

Differentiating by T and using the Gibbs-Helmholtz relation, we obtain

$$-\frac{\Delta\overline{H}_{\text{vap}}}{T^2} = \frac{d}{dT}(R\ln a_1) \qquad \textbf{(A-IV-8)}$$

Integrating between $T_b{}^1$, the boiling point of pure solvent (at which point $a_1 = 1$), and T_1 we have

$$-\int \frac{\Delta \overline{H}_{\text{vap}}}{\text{T}^2} d\text{T} = \int_{a_1=1}^{a_1} d(\text{R} \ln a_2) \qquad \textbf{(A-IV-9)}$$

If $\Delta \overline{H}_{\text{vap}}$ is assumed to be temperature independent, (A-IV-9) yields

$$-\frac{\Delta \overline{H}_{\text{vap}}(\text{T} - \text{T}_{\text{b}}{}^1)}{\text{RT T}_{\text{b}}{}^1} = \ln a_1 \qquad \textbf{(A-IV-10)}$$

At sufficiently small values of a_1, $a_1 = \text{N}_1 = 1 - \text{N}_2$, and $\ln(1 - \text{N}_2)$ is approximately (for $\text{N}_2 \ll 1$), $-\text{N}_2$. Equation (A-IV-10) therefore reduces to

$$\Delta \text{T}_{\text{b}} = \frac{\text{RT}_{\text{b}}\text{T}}{\Delta \overline{H}_{\text{vap}}} \text{N}_2 \cong \frac{\text{RT}_{\text{b}}{}^2\text{N}_2}{\Delta \overline{H}_{\text{vap}}} \qquad \textbf{(A-IV-11)}$$

or to

$$\Delta \text{T} = \text{K}_{\text{f}}\text{M}_2$$

where K_{f}, as shown in Chapter IV, is $\dfrac{\text{M}_1\text{RT}_{\text{b}}{}^2}{1000\,\Delta \overline{H}_{\text{vap}}}$

Freezing-Point Lowering

At equilibrium between a pure solid and a solution with the solid as one constituent,

$$\overline{G}^{\circ,S} = \overline{G}^{\circ,L} + \text{RT} \ln a_1 \qquad \textbf{(A-IV-12)}$$

Proceeding as in the boiling-point elevation discussion, we obtain

$$\frac{\Delta \overline{H}_{\text{fus}}\,\Delta \text{T}}{\text{RT}_{\text{f}}\text{T}} = -\ln a_1 \qquad \textbf{(A-IV-13)}$$

or, with the same approximations as used in deriving (A-IV-11) we obtain

$$\Delta \text{T}_{\text{f}} \cong \frac{\text{N}_2\text{RT}_{\text{f}}{}^2}{\Delta \overline{H}_{\text{fus}}} \cong \frac{\text{M}_1\text{RT}_{\text{f}}{}^2\text{M}_2}{1000\,\Delta \overline{H}_{\text{fus}}} = K_{\text{f}}\text{M}_2 \qquad \textbf{(A-IV-14)}$$

Index

INDEX

1 2 3 4 5 6 7 8 9 10